Advances in Artificial Intelligence Methods Applications in Industrial Control Systems

Advances in Artificial Intelligence Methods Applications in Industrial Control Systems

Editor

Emanuele Carpanzano

MDPI • Basel • Beijing • Wuhan • Barcelona • Belgrade • Manchester • Tokyo • Cluj • Tianjin

Editor
Emanuele Carpanzano
University of Applied Sciences and Arts of Southern Switzerland
Switzerland

Editorial Office
MDPI
St. Alban-Anlage 66
4052 Basel, Switzerland

This is a reprint of articles from the Special Issue published online in the open access journal *Applied Sciences* (ISSN 2076-3417) (available at: https://www.mdpi.com/journal/applsci/special_issues/ai_industrial_control_systems).

For citation purposes, cite each article independently as indicated on the article page online and as indicated below:

LastName, A.A.; LastName, B.B.; LastName, C.C. Article Title. *Journal Name* **Year**, *Volume Number*, Page Range.

ISBN 978-3-0365-6808-9 (Hbk)
ISBN 978-3-0365-6809-6 (PDF)

Cover image courtesy of Laboratory of Automation, Robotics and Machines (ARM-Lab) of the University of Applied Sciences and Arts of Southern Switzerland (SUPSI).

Contents

About the Editor

Emanuele Carpanzano

Emanuele Carpanzano is Director of Research, Development, and Knowledge Transfer at the University of Applied Sciences and Arts of Southern Switzerland. He has managed numerous R&D initiatives at international, national and regional level in addition to industrial research and knowledge transfer projects. He is active in different federal and international associations and institutions dedicated to education, research, and innovation programs and initiatives. At international level, he is member of CIRP (The International Academy for Production Engineering) and of EIT (European Institute of Technology) Manufacturing. At federal level, he is a member of SATW (Schweizerische Akademie der Technischen Wissenschaften), of the board of directors of FTAL (Fachkonferenz Technik, Architektur und Life Sciences), of Euresearch, and of the AM-TTC Alliance (Advanced Manufacturing Technology Transfer Centers Alliance). At regional level, he is a member of the boards of the Lifestyle Tech Competence Center and of SDBC, Swiss Drone Base Camp, both part of Innovation Park Ticino. As a scientific expert, he supervises, acting also as reviewer and independent scientist, research and innovation activities and projects at local, national, and European level. His research interests and activities are focused on industrial control and automation systems as well as on the digitalization of production systems and value chains, including evolution of related human aspects. He is Professor of industrial plants at SUPSI and author of more than 140 scientific papers as well as of different industrial patents in his applied research fields."

Preface to "Advances in Artificial Intelligence Methods Applications in Industrial Control Systems"

Artificial intelligence methods are being increasingly applied at different industrial control systems levels, from single automation devices up to the real-time control of complex machines, production processes, and overall factories in terms of supervision and optimization. Such innovative solutions are being exploited with reference to different industrial control applications from sensor fusion methods to novel model predictive control techniques, from self-optimizing machines to collaborative robots, and from factory adaptive automation systems to production supervisory control systems. The aim of this Special Issue is to provide an overview of novel applications of AI methods to industrial control systems, so as to improve the production systems self-learning capacities, their overall performance, the related process and product quality, the optimal use of resources, and industrial systems safety and resilience to varying boundary conditions and production requests.

Emanuele Carpanzano
Editor

applied sciences

Editorial

Editorial of the Special Issue "Advances in Artificial Intelligence Methods Applications in Industrial Control Systems"

Emanuele Carpanzano

Department of Innovative Technologies, University of Applied Sciences and Arts of Southern Switzerland, 6928 Manno, Switzerland; emanuele.carpanzano@supsi.ch

1. Motivation for the Special Issue

Today, Artificial Intelligence (AI) applications are considered to be of increasing relevance for the future of industrial control systems [1–3]. AI methods are increasingly being applied at different industrial control systems levels, from single automation devices up to the real-time control of complex machines, production processes and overall factories supervision and optimization.

AI solutions are exploited with reference to different industrial control applications, from sensor fusion methods to novel model predictive control techniques, from self-optimizing machines to collaborative robots, and from factory adaptive automation systems to production supervisory control systems [1,4–13].

The motivation for the present Special Issue is to provide an overview of novel applications of AI methods to industrial control systems by means of selected best practices, to highlight how such methodologies can be used to improve the production systems self-learning capacities, their overall performance, the related process and product quality, the optimal use of resources and the industrial systems safety, and resilience to varying boundary conditions and production requests [1,4,12,13].

2. Summary of the Special Issue Contents

This Special Issue includes seven papers: a perspective paper to frame the ongoing advances in AI methods applications in Industrial Control Systems to realize self-optimizing manufacturing systems [1], and six papers presenting novel applications of AI methods with reference to specific control problems and different application sectors [5–10]. Further details about the seven articles are provided in the following.

The perspective paper "Advances in Artificial Intelligence Methods Applications in Industrial Control Systems: Towards Cognitive Self-Optimizing Manufacturing Systems", by Emanuele Carpanzano and Daniel Knüttel [1], provides an overview of the recent progress on different hierarchic and functional levels of industrial control solutions, discussing how AI-based methods enable further enhancements in many applications and sectors. Promising algorithms and methods are presented and discussed for every control level following a bottom-up approach, starting from the sensor level and data fusion to more complex applications such as self-optimizing machines. It is outlined how the combination of advanced monitoring, modeling, control and scheduling methods will, in the future, allow for the development of self-optimizing machines, leading to production machine improvements in terms of product quality, productivity and resource efficiency, and representing a crucial point for the next generation of human-centric manufacturing systems.

The article "*Adaptive Optimal Robust Control for Uncertain Nonlinear Systems Using Neural Network Approximation in Policy Iteration*", by Dengguo Xu, Qinglin Wang and Yuan Li [5], presents an optimal adaptive control approach to solve the robust control problems of nonlinear systems with internal and input uncertainties, based on the policy iteration

Citation: Carpanzano, E. Editorial of the Special Issue "Advances in Artificial Intelligence Methods Applications in Industrial Control Systems". *Appl. Sci.* **2023**, *13*, 16. https://doi.org/10.3390/app13010016

Received: 14 December 2022
Accepted: 19 December 2022
Published: 20 December 2022

(PI) in reinforcement learning (RL). First, the robust control is converted into solving an optimal control containing a nominal or auxiliary system with a predefined performance index. It is demonstrated that the optimal control law enables the considered system to become globally asymptotically stable for all admissible uncertainties. Second, based on the Bellman optimality principle, online PI algorithms are proposed to calculate robust controllers for the matched and mismatched uncertain systems. The approximate structure of the robust control law is obtained by approximating the optimal cost function with the neural network in PI algorithms. Finally, to illustrate the availability of the proposed algorithm and theoretical results, some numerical examples are provided.

The paper "*Digital Twin for Designing and Reconfiguring Human–Robot Collaborative Assembly Lines*", by Niki Kousi, Christos Gkournelos, Sotiris Aivaliotis, Konstantinos Lotsaris, Angelos Christos Bavelos, Panagiotis Baris, George Michalos and Sotiris Makris [6], discusses the use of DTs (Digital Twins) for designing and redesigning flexible production systems. These systems employ mobile dual arm robots that move across the factory undertaking multiple tasks, assisting humans. Exploiting this hardware ability, the DT based re-design system generates optimal configurations in terms of layout and task plans. The solution allows for online reconfiguration of the system by (a) dynamically re-assigning the tasks when unexpected events occur, and (b) making real-time adjustments of robots' behavior to ensure collision-free trajectories generation in unstructured environments. These are achieved in addition to the sensor-based real-time scene reconstruction that is provided by the Digital Twin by synthesizing multiple sensor data. The discussed DT-based system was deployed, tested and validated in a case study from the automotive sector.

The contribution "Manufacturing Execution System Integration through the Standardization of a Common Service Model for Cyber-Physical Production Systems", by Richárd Beregi, Gianfranco Pedone, Borbála Háy and József Váncza [7], proposes a generalized common service model and architecture of CPS (Cyber Physical Systems)-based manufacturing execution systems. The core model is minimalist as far as its underlying assumptions are concerned. Hence, it does not constrain the decision autonomy of collaborating cyber-physical entities and "only" provides channels for transferring and synchronizing the information that ensues from their decisions. The proposed approach identified the elementary concepts, such as functions, calls, variables, and reports, as the basis for modeling and providing I4.0-compliant, CPS-based services in a manufacturing environment. They were developed via standardized technologies enabling semantic interoperability and openness (OPC UA, MQTT). The universal CPS-based service integration mechanism was validated in an experimental pilot production and logistics system that included a variety of heterogeneous and autonomous resources, such as manufacturing cells, AGVs, robots, and human–robot collaborative cells. These CPS components were connected and controlled via the plug and collaborate mechanism of a MES system in a number of complex scenarios.

The manuscript "*Decentralized Multi-Agent Control of a Manipulator in Continuous Task Learning*", by Asad Ali Shahid, Jorge Said Vidal Sesin, Damjan Pecioski, Francesco Braghin, Dario Piga and Loris Roveda [8], formulates a decentralized and multi-agent approach in the form of an RL (Reinforcement Learning) problem, demonstrating the possibility of decentralizing a single manipulator controller by applying multiple agents in learning continuous actions. The purpose of this paper was to compare the feasibility of decentralization for a single robot to show the modularity in the joints, as well as to make a comparison between the centralized and decentralized approach. The results show that it is possible to decentralize the control action on the robot. Using multiple agents shows the stagnation of the learning process when using the same reward function that is used for a single agent. It is believed that this is due to the lack of communication between the agents and the generality of the reward when considering two agents.

The paper "*Predictive Control for Small Unmanned Ground Vehicles* via *a Multi-Dimensional Taylor Network*", by Yuzhan Wu, Chenlong Li, Changshun Yuan, Meng Li and Hao Li [9], presents an improved predictive control scheme based on the MTN (multi-dimensional Taylor network) for tracking the control of SUGVs (Small Unmanned Ground Vehicles).

The traditional objective function was improved to obtain a predictive objective function with the differential term. The optimal control quantity was given in real time through iterative optimization. A tracking control experiment was carried out on an SUGV to verify the effectiveness of the proposed scheme. The results show that the proposed scheme is effective and has good a real-time, robustness, and convergence performance, which ensure that the vehicle can quickly and accurately track the desired yaw velocity signal and that it is superior to the traditional MTN and RBF (Radial Basis Function) predictive control schemes.

The article *"Design of a NARX-ANN-Based SP Controller for Control of an Irrigation Main Canal Pool"*, by Ybrain Hernandez-Lopez, Raul Rivas-Perez and Vicente Feliu-Batlle [10], proposes a modification of the well-known Smith predictor controller, in which the internal linear model was substituted by the combination of a NARX-ANN-based model and a TD-NARX-ANN-based model, in order to take into account the dynamic nonlinearities in the effective control of water distribution in an irrigation main canal pool. By the application of system identification procedures, NARX-ANN and a TD-NARX-ANN with recurrent architectures were obtained which describe with high accuracy the non-linear dynamic behavior of the water distribution in the studied canal pool. The NARX-ANN structure with an input layer with 13 memory blocks for the input and output signals, three neurons in the hidden layer and one neuron in the output layer, provided the best model performance. A third model—which was linear and represented by a time-delay first-order transfer function—was obtained using an identification procedure. The validation results of the three models illustrate that the FIT performance indexes of the NARX-ANN-based models are higher than that of the linear model.

3. Discussion and Concluding Remarks

By means of its seven scientific papers, the present Special Issue clearly illustrates the increasing added value of the introduction of AI methods to improve the performance of control solutions with reference to different control and automation problems in different industrial applications and sectors [1,4–10], ranging from single manipulators [8] or small unmanned ground vehicles [9], up to complex manufacturing plants [7] or irrigation systems [10]. Additionally, the role of AI to improve the performance of relevant engineering methodologies and digital instruments, such as Cyber Physical Systems [7], Digital Twins and Human-Robot Collaboration [6], are also effectively addressed in the contributions to the Special Issue.

Even if many research initiatives are ongoing, such as those discussed in the present Special Issue, many relevant challenges remain to be faced in the near future to exploit the high potential of AI methods for industrial control applications. In particular, it is of major importance to support the adoption of such novel solutions in real-world industrial applications, as well as by small- and medium-sized companies [2,3,11], and to integrate AI-based techniques to steer the co-evolution of human workers' tasks and manufacturing systems frameworks, in order to improve production systems' performance, as well as human workers' safety and well-being [12,13].

Funding: This research received no external funding.

Conflicts of Interest: The authors declare no conflict of interest.

References

1. Carpanzano, E.; Knüttel, D. Advances in Artificial Intelligence Methods Applications in Industrial Control Systems: Towards Cognitive Self-Optimizing Manufacturing Systems. *Appl. Sci.* **2022**, *12*, 10962. [CrossRef]
2. Carpanzano, E.; Jovane, F. Advanced Automation Solutions for Future Adaptive Factories. *CIRP Ann. Manuf. Technol.* **2007**, *56*, 435–438. [CrossRef]
3. Brusaferri, A.; Ballarino, A.; Carpanzano, E. Reconfigurable knowledge-based control solutions for responsive manufacturing systems. *Stud. Inform. Control.* **2011**, *20*, 32–41. [CrossRef]
4. Carpanzano, E.; Cesta, A.; Orlandini, A.; Rasconi, R.A. Valente, Intelligent dynamic part routing policies in Plug&Produce Reconfigurable Transportation Systems. *CIRP Ann. Manuf. Technol.* **2014**, *63*, 425–428. [CrossRef]

5. Xu, D.; Wang, Q.; Li, Y. Adaptive Optimal Robust Control for Uncertain Nonlinear Systems Using Neural Network Approximation in Policy Iteration. *Appl. Sci.* **2021**, *11*, 2312. [CrossRef]
6. Kousi, N.; Gkournelos, C.; Aivaliotis, S.; Lotsaris, K.; Bavelos, A.C.; Baris, P.; Michalos, G.; Makris, S. Digital Twin for Designing and Reconfiguring Human–Robot Collaborative Assembly Lines. *Appl. Sci.* **2021**, *11*, 4620. [CrossRef]
7. Beregi, R.; Pedone, G.; Háy, B.; Váncza, J. Manufacturing Execution System Integration through the Standardization of a Common Service Model for Cyber-Physical Production Systems. *Appl. Sci.* **2021**, *11*, 7581. [CrossRef]
8. Shahid, A.A.; Sesin, J.S.V.; Pecioski, D.; Braghin, F.; Piga, D.; Roveda, L. Decentralized Multi-Agent Control of a Manipulator in Continuous Task Learning. *Appl. Sci.* **2021**, *11*, 10227. [CrossRef]
9. Wu, Y.; Li, C.; Yuan, C.; Li, M.; Li, H. Predictive Control for Small Unmanned Ground Vehicles via a Multi-Dimensional Taylor Network. *Appl. Sci.* **2022**, *12*, 682. [CrossRef]
10. Hernandez-Lopez, Y.; Rivas-Perez, R.; Feliu-Batlle, V. Design of a NARX-ANN-Based SP Controller for Control of an Irrigation Main Canal Pool. *Appl. Sci.* **2022**, *12*, 9180. [CrossRef]
11. Bettoni, A.; Matteri, D.; Montini, E.; Gladysz, B.; Carpanzano, E. An AI adoption model for SMEs: A conceptual framework. *IFAC-Pap.* **2021**, *54*, 702–708. [CrossRef]
12. Bracco, F.; Bruzzone, A.A.; Carpanzano, E. Transfactory: Towards a New Technology-Human Manufacturing Co-evolution Framework. In *Advances in System-Integrated Intelligence*; Valle, M., Lehmhus, D., Gianoglio, C., Ragusa, E., Seminara, L., Bosse, S., Ibrahim, A., Thoben, K.-D., Eds.; Springer International Publishing: Cham, Switzerland, 2023; pp. 636–645.
13. Bettoni, A.; Montini, E.; Righi, M.; Villani, V.; Tsvetanov, R.; Borgia, S.; Secchi, C.; Carpanzano, E. Mutualistic and Adaptive Human-Machine Collaboration Based on Machine Learning in an Injection Moulding Manufacturing Line. In *Procedia CIRP—53rd CIRP Conference on Manufacturing Systems*; Elsevier: Amsterdam, The Netherlands, 2020; Volume 93, pp. 395–400. [CrossRef]

applied sciences

Perspective

Advances in Artificial Intelligence Methods Applications in Industrial Control Systems: Towards Cognitive Self-Optimizing Manufacturing Systems

Emanuele Carpanzano [1,*] and Daniel Knüttel [2,3]

1 SUPSI, University of Applied Sciences and Arts of Southern Switzerland, Via Pobiette 11, 6928 Manno, Switzerland
2 Intelligent Production Machines, Inspire AG, Via la Santa 1, 6962 Viganello, Switzerland
3 Institute for Machine Tools and Manufacturing (IWF), ETH Zürich, Leonhardstrasse 21, 8092 Zurich, Switzerland
* Correspondence: emanuele.carpanzano@supsi.ch

Abstract: Industrial control systems play a central role in today's manufacturing systems. Ongoing trends towards more flexibility and sustainability, while maintaining and improving production capacities and productivity, increase the complexity of production systems drastically. To cope with these challenges, advanced control algorithms and further developments are required. In recent years, developments in Artificial Intelligence (AI)-based methods have gained significantly attention and relevance in research and the industry for future industrial control systems. AI-based approaches are increasingly explored at various industrial control systems levels ranging from single automation devices to the real-time control of complex machines, production processes and overall factories supervision and optimization. Thereby, AI solutions are exploited with reference to different industrial control applications from sensor fusion methods to novel model predictive control techniques, from self-optimizing machines to collaborative robots, from factory adaptive automation systems to production supervisory control systems. The aim of the present perspective paper is to provide an overview of novel applications of AI methods to industrial control systems on different levels, so as to improve the production systems' self-learning capacities, their overall performance, the related process and product quality, the optimal use of resources and the industrial systems safety, and resilience to varying boundary conditions and production requests. Finally, major open challenges and future perspectives are addressed.

Keywords: control systems; industrial automation; artificial intelligence; machine learning; self-learning machine tools; adaptive production systems

Citation: Carpanzano, E.; Knüttel, D. Advances in Artificial Intelligence Methods Applications in Industrial Control Systems: Towards Cognitive Self-Optimizing Manufacturing Systems. *Appl. Sci.* **2022**, *12*, 10962. https://doi.org/10.3390/app122110962

Academic Editor: Alessandro Gasparetto

Received: 29 September 2022
Accepted: 24 October 2022
Published: 29 October 2022

1. Introduction

Currently, manufacturing companies have to deal with many challenges in order to remain competitive within the rapidly changing market dynamics and framework conditions, including socio-economic, environmental and cultural aspects.

Short delivery times and customization along with comparable production costs and high-quality parts are order-winning properties and therefore vital for business. Additionally, accelerated product developments and industrializations are important attributes to quickly cover market developments with brief time to markets. Moreover, upcoming regulations as well as growing social and environmental requirements represent additional challenges to be faced by the industry. Key enabling factors to face such challenges are production machines, whose developments are crucial to meet the cited and constantly growing demands of modern manufacturing companies.

To keep pace with these ongoing evolutions and master upcoming challenges, industrial control systems are a key factor for advanced production machines and industrial plants [1]. Closed loop and real time control solutions are already widely used and are well established on various levels of production systems. However, with increasing complexity of the manufacturing systems and control requirements, further developments of control algorithms are indispensable to address the demand for future manufacturing systems.

Promising approaches for improved industrial control systems are based on recent developments in artificial intelligence (AI). AI has obtained wide popularity in the past years and is one of the most pursued topics in research and industry. Progress in computing hardware and software has facilitated the collection of data and its exploration with AI for various use cases. Initiatives such as Industry 4.0 have further highlighted the potential of AI-based methods and have pushed them to the centre of attention in many fields [2]. A report from McKinsey & Company states that AI could contribute with a value of USD 3.6–6.6 trillion annually in the supply-chain management and manufacturing sector [3].

Especially in the area of manufacturing, AI-based developments further expand the capabilities and pave the way for more advanced industrial control systems. Current applications and trends are shown in the following section for various levels.

In addition to existing works about applications of AI or machine learning in manufacturing and Industry 4.0 [4–7], this work provides a perspective overview of the current state of the art, recent developments and future challenges for each manufacturing process and control level.

Therefore, this paper is structured following a hierarchical and functional bottom-up approach: starting from the processing and elaboration of sensor signals to the job scheduling operations of the whole production plant. Thereby, the focus lies on identifying and illustrating most promising artificial intelligence-based methods coping with current challenges at each industrial control and automation level, including the related data acquisition, process modelling, real-time control and task scheduling operations. Due to the plethora of existing algorithms, the aim is not to present a complete overview of existing approaches and methods, but to highlight the most promising ones and the potential of their future development and applications to face modern industrial control and automation challenges, by means of emerging AI-based methodologies. In particular, the key role of the integration of AI methods in industrial control solutions is to strive for future self-optimizing production systems. For the successful implementation of self-optimizing machines, all components over all hierarchies must be optimally designed and operational. Figure 1 schematically illustrates the main element of self-optimizing production systems and forms the basis for the focus of the first sections in the present review and survey paper. Starting from a controller dedicated to a set of well-defined process variables, multiple aspects must be considered to enable an accurate and reliable control that is elementary for subsequent higher-level control structures. The first step of each controller is the direct or indirect measurement of the process variable under consideration. In order to obtain the most relevant information and controller inputs, sensor signals must be processed and elaborated accordingly. Since amount, type and quality of sensor signals can vary significantly, different methods are required to process them in an appropriate manner. In particular, trends towards the utilization of multiple sensors and the consequently increasing data availability require new developments towards suitable approaches for sensor and data fusion. Related current approaches and AI-based methods are discussed in Section 2.

A further important element within optimized controllers and manufacturing applications are process models. Depending on the use case, these process models can be generated conventionally with multi-physics simulations as finite element methods or more recently, data-driven approaches enable the generation of process models based on the previously acquired data. However, not all process models are suitable for control applications. An overview of available approaches and recent developments concerning AI-based process models is given in Section 3. By processing the measurement values and querying process models, the optimizer determines the optimal controller output. Depending on the application, these processes must be performed very fast or even in real time and with as much autonomy as possible, requiring intelligent and specialized solutions. Therefore, promising AI-based methods to implement effective optimizers are presented in Section 4.

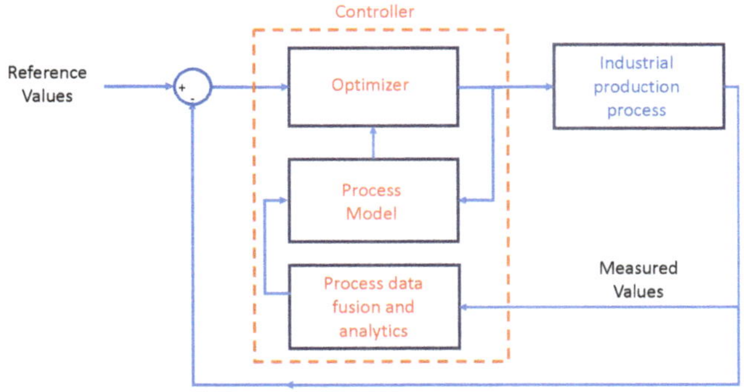

Figure 1. Model-based optimized controller.

A further concatenation of multiple optimal controllers enables an advanced control strategy over all hierarchical process levels. The main elements of self-optimizing machines are their capability to autonomously adapt to changing requirements and circumstances. Therefore, the task scheduler coupled with multiple cascaded control chains allows one to redefine and modify control objectives of control chains towards higher-level optimal objectives. Thereby, an example for the cascaded control chains could be the process control of an additive manufacturing process, where the first controller guarantees a constant energy density during the deposition process and the second controller acts layerwise by modulating the process parameter for the subsequent layer to obtain an equal deposition height. In an exemplary additive manufacturing process, the two control loops could represent the regulation of two laser sources, simultaneously building various features of a more complex part. A simplified overview of a self-optimizing framework is displayed in Figure 2.

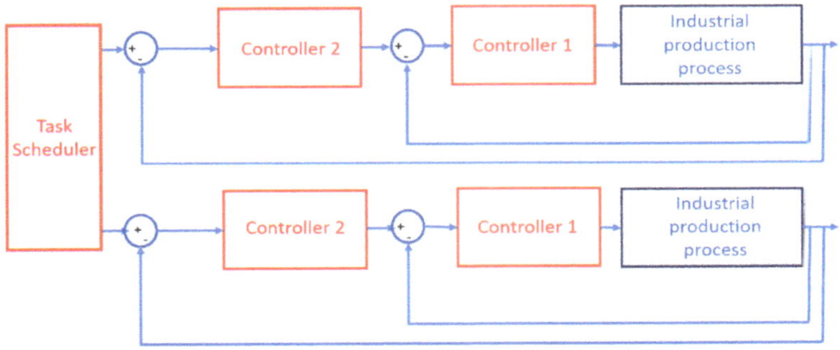

Figure 2. Self-optimizing production system control framework.

AI methods can enhance the self-optimization capacity of a cognitive industrial control system at different hierarchical and functional levels. The combination of models and real-time data, as well as of real-time control and job scheduling functionalities, allows for the development of self-optimizing industrial machines and systems with new elevated degree of autonomy. Current methods and developments of job scheduling functionalities are described in Section 5, whereas approaches and frameworks towards cognitive self-optimizing machines are outlined in Section 6.

2. Methods for Industrial Data Fusion at Sensor, Feature and Decision Level

Efficiency, repeatability, robustness, and quality of manufacturing processes are of great importance in today's production sites. Monitoring systems enable to capture parameters and system states in real time, allowing to effectively control manufacturing processes. The quality and performance of these automation systems stand and fall with the reliability and precision of available information. As a result, robust and accurate process measurements are of crucial relevance to satisfy higher-quality requirements.

A common approach to tackle these issues is a fast developing and huge research field called sensor fusion. The scope of data fusion, also known as sensor fusion, multi-sensor data fusion, multi-sensor integration, data combination or data aggregation is to obtain improved information with higher accuracy by merging multiple information sources. It fosters the robustness, reliability and confidence of the analysed information and improves the temporal and spatial resolution [8]. Initial developments of sensor-fusion-based systems go back to the military industry, where the authentication and tracking of dynamic objects is of crucial importance. Recently, further sectors such as the automotive sector experience a strong increase in sensor fusion for autonomous vehicles to accurately map their environment.

There have been multiple attempts to define a definition and framework for sensor data fusion. One of the most popular definitions has been presented by the Joint Directors of Laboratories (JDL) [9] describing sensor fusion as a multi-level process handling the association, correlation and combination of data and information from single and multiple sources. This section focuses on the application of AI-based algorithms for industrial data fusion. For a deeper insight in further data fusion architectures, the reader is referred to [10–12].

To capitalize on the benefits of sensor data fusion, some challenges must be overcome for a successful implementation. There are various data properties affecting the feasibility and success of the data fusion process. The main challenges of data fusion arise from the heterogeneity of data sources and individual imperfections of each data source such as uncertainty, imprecision or granularity. An overview of typical data fusion issues is displayed in Figure 3.

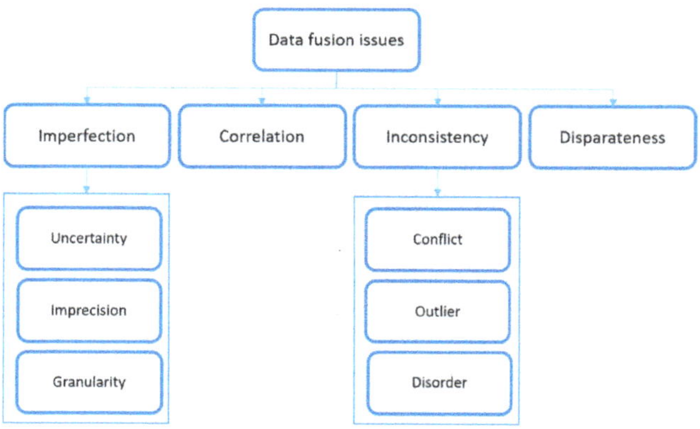

Figure 3. Taxonomy describing data-fusion-related issues.

Depending on the considered data type and its fusion issues, suitable fusion algorithms must be selected. Today's available algorithms only address a subset of these aspects. Due to the multitude of applications and available algorithms, the following section does not claim to be complete, but identifies promising methodologies for applications in industrial control solutions. Furthermore, there are many ways to classify fusion algorithms. Meng et al. [13] presented the algorithms according to their abstraction level as signal level, feature level and decision level, whereas Sasiadek et al. [14] divides fusion approaches in probabilistic models, least-square techniques and intelligent fusion.

On a *signal level*, the most intuitive and easiest way to increase the signal-to-noise ratio of redundant sources is to apply the weighted average or the least squares method on captured data. A more sophisticated and widespread method is the Kalman Filter. The Kalman Filter considers the statistical characteristics of the measurements while still allowing for real-time processing of the data. Kalman filtering is a discrete and time invariant approach and only holds for linear models. Unfortunately, practical problems often have strong non-linear relations. To overcome these limitations, Extended Kalman Filter (EKF) [15], Unscented Kalman Filter (UKF) or the Particle Filter (PF) [16] also known as Sequential Monte Carlo (SMC) have been developed to handle non-linear relationships. A drawback of the latter method is the increasing computational effort and rising amount of required data.

Regarding the *feature level*, unsupervised learning algorithms such as K-means clustering or self-organizing maps can be used to classify relationships between entities and extract further information. These methods are particularly suitable for applications where patterns or geometrical relations need to be detected. Further methods to address pattern recognition problems are adaptive resonance theory (ART), ARTMAP or Fuzzy ART.

For applications with limited bandwidth, Challa et al. [17] applied support vector machines (SVM) to compress the information in sensor fusion for large amounts of samples. Further algorithms and their applications are presented by Fung et al. [18]. Depending on the use case and its requirements, the suitability of the data fusion algorithm must be evaluated individually. Thereby, the target application defines the required accuracy, computational complexity, processing power and available amount of data.

King et al. [16] introduced a classification depending on the degree of complexity. Thereby, low-complexity algorithms such as k-means allow for an enhanced battery life for wearable health monitoring systems due to their reduced computational complexity and resource requirements.

In Table 1, the presented algorithms are distinguished regarding their application or abstraction level as sensor fusion, feature fusion and decision fusion. In general, there is a wide range of possible data fusion applications.

Table 1. Overview of data fusion algorithms at sensor, feature and decision level.

Abstraction Level	Method	Applications/References
Sensor Level	Weighted average	[19]
	Kalman Filter (EKF, UKF)	[20,21]
	Particle Filtering	[22,23]
	Dempster-Shafer method	[24,25]
Feature level	k-nearest neighbour (k-NN), k-means	[26]
	Decision Trees	[27]
	Support vector machines (SVM)	[28]
	Artificial neural networks (ANN)	[29–31]
	Gaussian mixture model (GMM)	[32,33]
Decision level	Bayesian inference	[34,35]
	Fuzzy logic	[36,37]

Typical examples of the military industry are the location and tracking of moving objects. Depending on the target, multiple sensor platforms are used, and various sources as acoustic signals or electromagnetic radiation can be observed [38]. More recent applications are in the health care [39] or telecommunication sectors as wireless sensor networks. The widespread machine learning algorithms such as ANN, SVM or GMM lead to an increased adoption in sensor fusion approaches and enhance the possibilities of sensor fusion.

However, the efficient handling of big amounts of data in industrial applications remains an important challenge for the future. Approaches called consensus filters try to address this issue by implementing distributed filters to create scalable algorithms [40,41]. More recent developments were published in the automotive sector, where fast developments are noticeable. Future self-driving or autonomous vehicles require reliable information about their environment and are therefore predestined for sensor fusion [42–44]. Wei et al. [45] fusioned LiDAR and a vision system to reduce the false positive detection of the LiDAR system utilizing neural networks to project the camera to the LiDAR space. A more detailed review in this sector is presented by Wang et al. [42] and Alatise et al. [44].

These developments will probably lead to further adoption and improvements in manufacturing industries. Besides data-related issues, the interaction and data fusion of human inputs and machine data will play an important role in the future and requires more research to be developed in the near future.

3. Empirical Data Driven Methods for Production Process Modelling

Production process modelling enables the development of stable and repeatable process models, which are of crucial relevance for manufacturing companies. Production processes can be viewed from different perspectives ranging from the overall production process chain including raw material delivery, manufacturing steps and the final stock exploitation with the final product to simulations of the microscopic material behaviour during a manufacturing step [46]. This section focuses on the modelling of manufacturing processes for control purposes. The following sections consider high-level perspectives accounting for multiple manufacturing machines and systems.

In the domain of manufacturing processes, as for example additive manufacturing, models can be classified in analytical, numerical or empirical approaches [47]. Analytical models consist of the mathematical analysis of a physical problem described by partial differential equations (PDE) and given boundary conditions. Given the assumptions made, the obtained solution accurately represents the reflected domain. However, the process to determine analytical solutions is not trivial, and for complex processes, advanced numerical methods and demanding computational efforts are necessary to solve the considered physical equations.

Numerical methods are typically finite element method (FEM), boundary element method (BEM), finite difference method (FDM) or finite volume method (FVM). The solving of PDE is an essential element for multi-physics simulations enabling the simulation of complex manufacturing processes. Offline process simulations allow to perform parameter or design engineering without performing costly experiments and are therefore widely spread in R&D departments. An important limitation of these methods is the computational burden for complex processes including multiple physical phenomena. Furthermore, the computational effort scales drastically with increasing simulation domain sizes. Besides computational drawbacks, setting up a model is not trivial and requires highly skilled employees to define model parameters. The model definition and its assumptions are crucial to obtain representative results for a predefined process window.

In certain applications, it can be more suitable to create empirical models by performing a series of experiments and analysing the observations. Thus, relations between process inputs and obtained measurements can be captured without knowing the detail about the physical phenomena. This method is in particular suitable for low-cost experiments, where numerical or analytical models would require more resources. To approximate the captured observations in terms of an exploitable model, various approaches exist. One of the most straightforward methods in statistical analysis is linear regression, where a line is fitted to capture the relation between two variables. For more complex non-linear applications, machine learning based methods as neural networks (NN) have experienced a widespread adoption in recent years [48–53]. Due to the increased availability of computational resources, the increasing development of machine learning libraries and the possibility to approximate highly non-linear relationships lead to many applications in industry, as for example machining operations and additive manufacturing [54]. The approximation capacities can be customized by varying NN hyperparameters as the number of hidden layers or the number of neurons of each layer. A further advantage of the NN-based model is the possibility to apply the model for similar processes. Therefore, only few additional data points are required to fine tune the already trained model by adapting the model to the changed conditions [55]. However, the main drawbacks of NN are the required amount of data to train the network and the reliability of the predictions. The structure of NNs makes it difficult to interpret and predict the behaviour, in particular for unseen data points.

Recent developments combine the advantages of physics-based simulations and neural networks by including the physical knowledge in the training of the neural network [56–59]. In contrast to conventional neural networks, physics-informed neural networks contain additional loss terms. Besides the prediction error, terms satisfying the partial differential equation of the simulated domain penalize solutions disagreeing with the physical equations. As a result, less data are required to train and optimize the weights of the network, and predictions are more stable. Moreover, the additional loss terms shrink the solution space and prevent overfitting of the model. For data scare applications, this method is very promising but needs further research for industrial applications. In particular, the scalability needs to be investigated since the additional computations for each iteration and the coordinates dependent on PDE are potentially limiting factors for an adoption on larger part dimensions.

With a similar objective, but without directly implementing the physics of the considered process, a recent method has been presented by Li et al. [60]. The authors presented the Fourier neural operator (FNO) to learn the underlying multiple partial differential equations (PDE) by mapping from a functional parametric dependence to the solution. The proposed architecture method appears to be significantly faster than traditional PDE solvers. To train the FNO, a sequence of timeframes of a physical phenomenon is given as input to predict the following time frames. The inputs are processed by transforming the input into the frequency space, filtering the higher frequency modes and transforming the filtered data back to the time space; furthermore, the inputs are multiplied with weights and summed up with the filtered results. Subsequently, the results are passed through an activation function, and the described layer is repeated multiple times. The advantages of the method are the mesh invariance and the capability of learning the relevant physics behind the process. Therefore, this method could be a promising approach for the modelling of manufacturing processes. First adoptions for solving the PDE for two-phase flow fields have been demonstrated by Zhang et al. [61]. However, the transfer from theoretical physical problems to empirical data is still missing.

Besides AI-based algorithms inspired by the structure of NN, probabilistic approaches as Bayesian Optimization (BO) have also been proven to be suitable for data scare applications. Due to the probabilistic nature of BO, probabilities and uncertainties of the prediction are computed. These values enable the development of data sampling methods, minimizing the uncertainty of the predictions. As a result, less experimental data are required to obtain a model of the manufacturing process. Maier et al. [62] successfully applied this method to optimize grinding parameters after a few trials for an unknown workpiece and tool combination. Table 2 summarizes the mentioned process modelling methods.

However, a general challenge of process simulations remains the lack of generalizability of models and the increased computational burden for complex processes. Simulation domains are often tiny volumes and only represent a small element of the manufacturing step. Previously mentioned recent developments could be promising to handle both larger simulation domains and complex processes. Particularly, data-driven methods have recently shown great potential. Further research regarding the explainability and generalizability of data-driven methods must be carried out to enable an imminent transition to industry.

Table 2. Overview of sensor fusion algorithms.

Modelling Type	Modelling Methods	Applications
Analytical	Analytical	Well-known and simpler models
Numerical	FEM, FVM	Complex processes, limited domain size and computational expensive
Empirical/Data-driven	Statistical analysis, e.g., linear regression	Well-known, low computational cost, for less complex problems
	Neural network	Mapping of nearly arbitrary relations, with sufficient experimental data
	Gaussian processes	Process parameter optimization, difficulties in handling high-dimensional problems, small amounts of available data
	Physics-Informed Neural Networks	Process modelling, small amounts of data
	Fourier Neural Operator	Mesh invariant modelling methods. Novel approach needs to be tested for real use cases

4. Machine Learning Applications for Production Machines Real Time Control

Real-time control of manufacturing machines is a further essential element of production machines to achieve repeatable and stable processes with high-quality parts. Applications range from accurate positioning of machine axes to adaptive control of process parameters as the laser power during laser welding. Today's industrial control systems are mainly based on highly consolidated solutions, as for example PID controllers for axis control. The simple structure, ease of implementation and long experience in parameter tuning lead to an established technology for control applications. However, there were many further developments in the controller design and automated parameter tuning [63]. AI-based methods have been increasingly adopted in recent years to optimize control parameters tuning or to develop new types of control methods.

In general, machine learning methods can be divided into supervised, unsupervised and reinforcement learning (RL) [64]. Supervised learning trains for example an NN by reducing the mean squared error between predictions and the corresponding labelled solution. In unsupervised learning, no information about the ground truth is available. Unsupervised learning often consists in clustering or reducing data dimensionality by grouping correlated data points. In contrast to supervised and unsupervised learning, RL forms a semi-supervised learning approach, where an agent interacts with the environment and a reward is given for performed actions. The algorithms explore the environment by executing actions and receiving rewards for the actions and decisions taken. The optimal execution order of the actions, the policy, is not known and will be determined during the learning process. A good introduction into Deep RL is given by Mousavi et al. [65]. RL can be further subdivided into dynamic programming, Monte Carlo methods and temporal difference methods [66].

Regarding PID parameter tuning, RL has been successfully applied to find suitable controller parameters. Besides traditional approaches such as trial and error or Ziegler–Nichols [67], Wang et al. [68] utilized an actor critic RL approach to determine the PID parameters. The results demonstrated the algorithm's ability to track non-linear systems while ensuring adaptability and robustness. Besides parameter tuning, RL also experiences an increased adoption within process control applications. Table 3 lists the currently most adopted RL algorithms for control applications.

Table 3. Overview of common and new control algorithms.

Type	Control Method
Traditional Fuzzy Logic	PID, fractional PID Fuzzy controller
Reinforcement Learning	Temporal difference (TD), SARSA (State–Action–Reward–State–Action)
	Soft Actor–Critic (SAC) or Asynchronous Advantage Actor–Critic (A3C)
	Q-learning, Double Q-learning

Nian et al. [66], Shin et al. [63] and Lee et al. [69] recently published reviews of machine learning and process control methodologies. In particular, Shin et al. [63] compared reinforcement learning with model predictive control algorithms. For complex and dynamical systems, model predictive controllers (MPC) represent an effective alternative to PID controllers. Due to their ability to explicitly consider constraints and predictively optimize the controller output, MPC solutions are already well established in the industry. Important for the applicability of MPC are sufficient slow system dynamics so as to execute the model online during every sampling step. Reinforcement learning for process control appears to be a promising method to reduce online computation times. The m ain differences lie in the modelling approach, where RL algorithms learn the model through trial and error, whereas MPC requires a model developed with first principles or process identifications. Furthermore, the online execution time of model-free RL is faster than MPC, since only a forward run of the policy network is required, whereas the MPC requires solving an optimization problem at each execution. A general comparison between deep RL and MPC for optimal control is given by Lin et al. [70] and is displayed in Table 4. More details about the comparison of MPC and RL can be found in [70,71].

Table 4. Comparison of deep reinforcement learning and model predictive control for optimal control [72].

Comparison Criteria	Reinforcement Learning	MPC
Solution form	Neural networks	Model based optimization
Online computation time	Low	High, especially for high-degree non-linear systems
Optimality-seeking capability	Near-optimal	Near-optimal with long prediction horizons
Generalization issue of machine learning	Reduced optimality	N/A
Handling of hard constraints	Under development	Yes
Handling of modelling errors, control delays, and/ or disturbances	Better with larger errors	Worse if not robust MPC

Due to the possibility of offline training, reinforcement learning is suitable for real-time control. Famous results of reinforcement learning were achieved by DeepMind beating human players in games such as Go [72]. However, these kinds of simulated environments are perfectly known and are therefore particularly suitable for the learning of reinforcement algorithms. In contrast to simulated environments, industrial processes suffer from high levels of imprecision and uncertainty (see sensor fusion; Section 2). The training requires a lot of data and interactions with the environment, which can be performed by interacting with a model. However, accurate models covering the entire manufacturing process over all parameter combinations are extremely complex, and experiments are costly. As discussed in Section 3, AI-based modelling approaches could therefore be suitable to model the process and serve as an environment for the RL algorithm. In contrast to traditional control approaches, there are no stability guarantees for RL controllers. Compared to traditional control methods, the development of RL control algorithms is still in an early stage. Nevertheless, the general learning capability of RL is promising for the future. Furthermore, research is needed to search for methods to improve stability and convergence of RL algorithms. For further applications as control approaches, safe RL will be, in the future, an important topic to enhance the reliability of RL-based controllers [73,74].

5. Real-Time Scheduling Methods for Flexible Manufacturing Systems

Trends towards customized products and shorter delivery times lead to a major challenge for manufacturing factories automation and management systems. The increase in product variants introduces significant non-production times due to system reconfiguration and set up needs at each product changeover. Besides internal production uncertainties, external factors such as a varying customer demand, supply chain ruptures or even raw material price dynamics increase the complexity level of the manufacturing configuration, planning and scheduling tasks [75]. To cope with these challenges, today's manufacturing companies need not only high-performance and flexible processes but also robust concepts to handle foreseen variances and unforeseen disturbances with minimal impact and small additional production costs [76,77]. Approaches to raise the production flexibility can range from improvements and development of adaptive work cells with self-adaptive control functions [78,79] to optimized production planning and scheduling solutions [80].

The mapping and optimization of these increasingly complex production processes is an important field of research. Recent works underline the relevance of production scheduling and show a tendency towards the following topics in the last years [81]:

- Self-organization of resources;
- Self-regulation of the production process;
- Self-learning capacities of the overall manufacturing industrial system.

These topics highlight the importance of autonomous and self-optimizing scheduling approaches to handle the overall growing production dynamics. Frequently applied scheduling optimization objectives are the maximization of machine utilization and the minimization of the lead time to produce a product [82]. Additionally, an increased sensibility and awareness of the climatic impact lead to further requirements such as the minimization of utilized resources and the reduced emission of greenhouse gases.

Once the scheduling objectives are defined, various approaches exist to implement a job shop scheduling policy. A general classification divides them into centralized, distributed or more recent hybrid scheduling and control architectures (HCA) [83]. These hybrid structures aim to combine the advantages of both approaches, while reducing their drawbacks. State-of-the-art architectures are multi-agent systems (MAS), holonic manufacturing systems (HMS) or product-driven systems (PDS) [83]. Another classification presented by Zhang et al. [84] groups the job shop scheduling models by the problem space in the following five types: basic type job shop scheduling problem (JSP), flexible JSP, multi-resources FJSP model (MrFJSP), multi-plants-based MrFJSP (MpFJSP) and MpFJSP with smart factory.

Independently of the scheduling architecture or model, dynamic scheduling approaches are required to cover and react on production events in real time. Events can be classified in resource- or job-related events [85]. Thereby resource-related events include process-linked incidents, as for example machine breakdowns or material shortages. Examples for job-related events are job cancellation or a change in priorities. Handling these dynamic and unforeseen events, while considering constraints and variations of each production process, is a key challenge of industrial production scheduling solutions. To cope with the challenge, accurate production models are required to obtain realistic results. However, with increasing model complexity, computation times and resulting optimization times increase significantly, limiting the real-time application potential of advanced scheduling methods. To overcome these issues, new technologies and algorithms are required to model and optimize the industrial operation scheduling problem.

Promising approaches are dynamic HCA with self-adaptive mechanisms to improve the agility and reactivity [86]. Cardin et al. [75] presented an optimized reactive control architecture (ORCA) consisting of a global and local control layer. Important is the distinction between normal and disrupted mode. In the case of a local perturbation, the local optimizer triggers the switch to disrupted mode, where the local optimizer controls the entity without the input of the global control layer. As a result, the system is able to handle reactive and local disturbances, while maintaining a global optimal solution at the start of each time window. The overall architecture is shown in Figure 4, illustrating the different control levels and the disrupted mode.

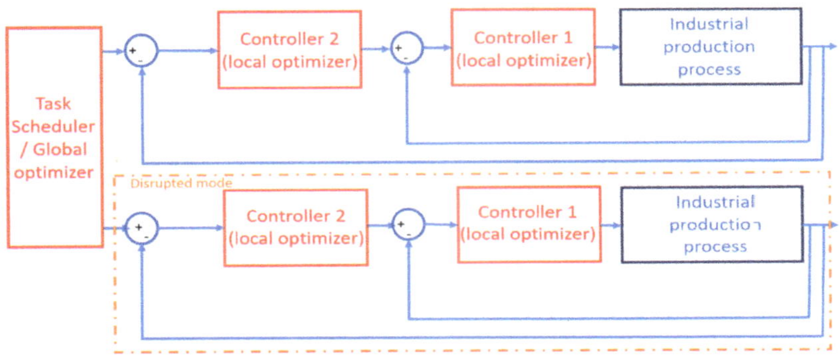

Figure 4. Optimized reactive control architecture (ORCA) [75].

However, a crucial element for effective industrial scheduling solutions is an algorithm that optimizes the objective functions in dynamic rescheduling operations. An overview of scheduling algorithms is shown in Figure 5 [84]. With increasing complexity, approximate methods are more suitable to optimize the production scheduling problem.

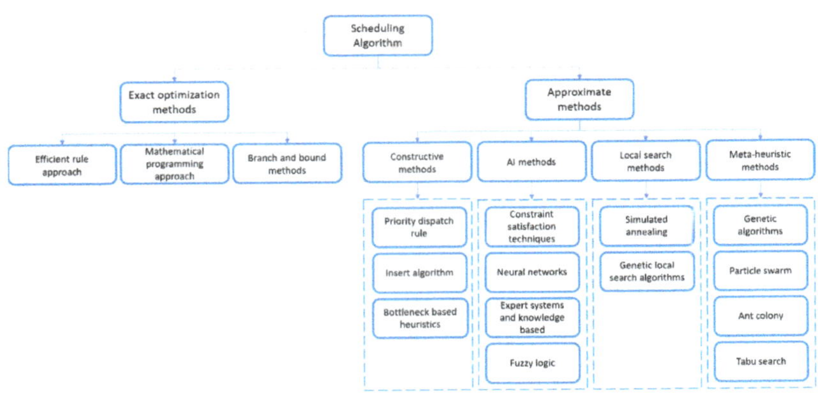

Figure 5. Scheduling algorithms classification.

In recent years, AI-based methods have received increasing attention and have been more commonly applied to support the scheduling process in high-performing industrial applications. Predictive maintenance enables to predict and even anticipate production disturbances. The research in this field has increased drastically due to the exponentially increasing amount of available production data and the adoption of AI-based techniques [87,88]. Developments in this area are especially helpful for scheduling solutions to anticipate disturbances and future changes, so as to proactively adapt the scheduling solution. Within the developments under Industry 4.0, new frameworks such as the MpFJSP with smart factory are presented [84]. Thereby the actual job shop situation can be measured in real time by various sensors and identification technologies. As a result, the actual situation can be compared to the planned scheduling solution. AI-based methods such as deep learning technologies enable to extract information and learn from the production processes autonomously to provide the necessary basis for decision making and new scheduling solutions identification.

Priore et al. [89] presented an approach to predict the suitable decision rule with machine learning based methods. Thereby, ensemble methods have been successfully applied to exploit the prediction of multiple algorithms and increase the prediction accuracy. Another work using ensemble methods has been presented by Lubosch et al. [90]. Thereby, a Monte Carlo Tree Search was used to develop a flexible scheduling algorithm. To achieve a good tradeoff between exploration and exploitation, the algorithms' hyper parameters are tuned with a Gradient Boosted Decision Tree (GBDT), avoiding a manual parameter tuning. Recently, reinforcement learning-based production scheduling systems have gained increasing attention in research [91]. Thereby a clear research trend towards more reliability-aware algorithms can be identified.

Besides further research in modelling and optimization algorithms, human aspects need to be considered in future adaptive automation solutions as well [92]. In particular, human–robot collaboration is a promising approach for higher productivity and product quality [93,94]. Worker-aware adaptive solutions enable the possibility to harmonize human well-being to obtain a synergic human-automation solution, including the related job scheduling aspects. Bettoni et al. [95] and Valente et al. [96] demonstrated an approach of adaptive huma–machine collaboration, enabling an effective reduction of the physical and mental stress of the operator. The monitoring of physiological parameters and classification with random forest algorithms allow for estimating the state of the operator and adapting the behaviour of the human-aware manufacturing systems, also considering the system evolution over time, as outlined in Bracco et al. [97].

6. Emerging Cognitive Approaches for Self-Optimizing Machines

The key idea behind self-optimizing machines (SOM) or self-optimizing production systems (SOPS) is to enable industrial production systems to produce customized high-quality parts while satisfying productivity demands [98]. Therefore, highly autonomous production machines are required to maximize the counteracting targets of efficient production and customization. Enhanced autonomy enables the system to produce complex parts, while handling and reacting to unforeseen events autonomously without any intervention of external operators [99].

Adapting the machine state automatically towards the objectives requires increased decision-making capabilities [100]. Thereby, the decision-making process can be compared to the human decision-making requirements [101]. These cognitive processes consist of the perception of the environment, its interpretation, the crosslink with existing knowledge and the subsequent decision with a coupled action. The general cognitive process is displayed in Figure 6 and illustrates the high-level control strategy.

The illustrated control structure includes the elements of each control loop. Cascading these process control loops forms a fundamental element to achieve this elevated degree of autonomy. These process control loops enable the SOM to react on changing objectives and subsequently adapt the machine state levelwise top–down. The cascaded implementation allows one to act on various influences on different impact levels, ranging from high-level product modifications to microscopic tool wear during the manufacturing process. Figure 7 visualizes a cascaded self-optimization loop for an additive manufacturing process. Thereby, the control loops are linked top–down to consider production objectives within the process control and act on each process level. Based on the Viable System model of Stafford Beer, Permin et al. [102] presented a concept of the Viable System model, including different levels of production and production management. Thereby, objectives are divided into normative production management, strategic production management and tactical production management.

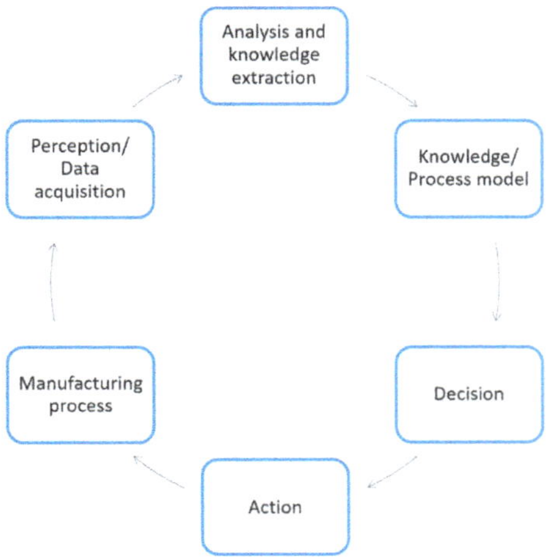

Figure 6. The cognitive process supporting self-optimization.

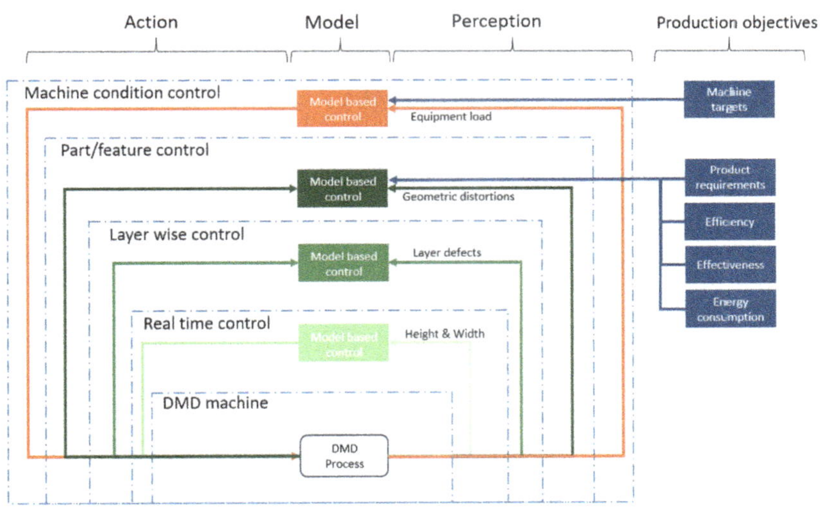

Figure 7. An exemplary cascaded loop of a self-optimizing additive manufacturing machine.

A main element of the SOM and its cascaded process loops is the exploitation of multiple information sources to capture and control the machine state. Moreover, the link between all processing elements allows for the process control over multiple levels. Thereby, process models as described in Section 3 support the control process by predicting target parameters or by optimizing control outputs with model-based controllers. Thombandsen et al. [103] presented a framework of a model-based self-optimizing manufacturing machine. Within this concept, accurate observations and the understanding of their relation to the product quality are crucial prerequisites for self-optimizing machines. Based on sensor signals, machine states and operating points are identified, and control parameters are computed by a model-based optimization.

A more higher-level perspective of a SOM is presented by Möhring et al. [104] and is displayed according to the framework in Figure 8. Within this concept, the process starts with generation of the trajectory for the imported part to manufacture. During this process, the planning process considers already processed and machine properties to optimize the toolpath. Further optimizations of the settings are performed by modelling and simulation of the process and the machine. These models enable the prediction of various product target properties, as for example the surface quality or part geometry. Afterwards, the actual machining process is performed, and monitoring solutions record the real-time process behaviour and interact with real-time controllers. The actual process measurements allow one to automatically update and improve the process model. Furthermore, the inline measurements form the input of real-time controllers to perform accurate control tasks, such as axis positioning. Considering the machine characteristics and measured process deviations, real-time controllers are constantly refined. Further details and examples can be found in Möhring et al. [104].

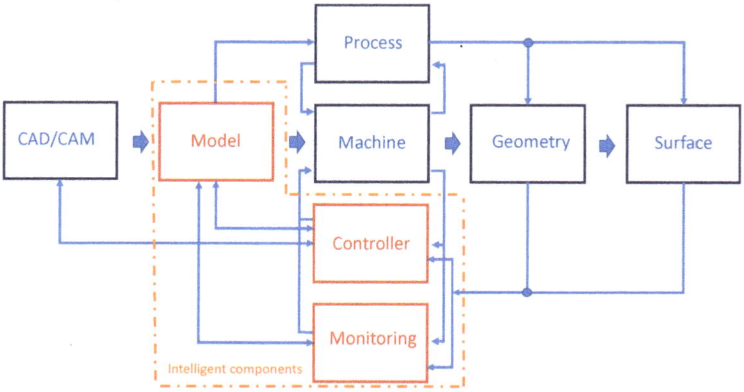

Figure 8. Self-optimizing manufacturing machine [104].

Essential prerequisites for SOM are developments in data fusion, process modelling and real-time control mentioned in the previous sections. In particular, the implementation of suitable measurement devices is crucial for the subsequent process steps. Only if the machine state is captured properly by accurate and robust measurements can valid process models be generated and subsequent optimal actions selected. Key enabling technologies as artificial intelligence and machine learning are promising approaches to process and analyse sensor data. These techniques are necessary to build cascaded process control loops based on comprehensive process observations, accurate process models and the knowledge about quality-influencing factors. Even if there have been promising advancements in various applications, there is still a lack in industrial implementations of comprehensive SOM solutions. A general overview of AI applications in manufacturing is given by Monostori [105]. Mayr et al. [106] presented an adaptive self-learning control approach to compensate for thermal errors on five-axis machine tools. Based on data captured during the process, a thermal autoregressive with exogenous input (ARX) model is identified and updated during the process to reduce thermal errors. Dittrich et al. [107] presented a self-optimizing process planning approach for five-axis milling to automatically compensate for tool deflections. Combining a simulation model with support vector regression enabled the prediction of shape errors. The subsequent automated toolpath adaption resulted in a shape error reduction by 50%.

The overall promising benefits of SOM are becoming more and more realistic with recent progresses in computational power, sensor techniques and AI algorithms. Autonomous selection of data sampling points enables an efficient generation of process models [64]. Based on these models, integrated optimization processes pave the way for autonomous self-tuning of parameters to guarantee a constant product quality.

To enable further widespread in industrial applications, faster and data-scarce simulation and analysis methods will be needed. AI algorithms are promising techniques to generate fast models and to handle process uncertainty, while updating the model during manufacturing processes. However, these approaches suffer from low data efficiency and reliability of its predictions and are therefore still critical regarding safety requirements.

Further challenges are the process dynamics, requiring fast optimized control values to control the process. Further developments are needed to cope with the complexity and dimensionality of the optimization problems.

Moreover, more research will be needed to integrate the human interaction and transparency of the decisions made and an interactive machine user interface.

7. Major Ongoing Research Directions and Future Perspectives

The presented work introduced recent progress on different hierarchic and functional levels of industrial control solutions, discussing how AI-based methods enable further enhancements in many applications and sectors. Promising algorithms and methods were mentioned and discussed for every level following a bottom-up approach, starting from the sensor level and data fusion to more complex applications such as self-optimizing machines.

Thereby, the (self)-learning properties of AI-based algorithms are promising approaches to deal with complex processes. Compared to conventional FEM-based modelling approaches, AI-based methods come with short inference times, which are important for real-time production systems. However, future research is required to overcome the drawbacks of AI-based methods as far as the model reliability and explainability are concerned. Approaches such as neural networks are difficult to interpret for data points outside the training set. A linked issue is the required amount of data for the training process, which can be a burden for data scare applications. Furthermore, data quality in terms of uncertainty, outliers or missing data points is crucial for accurate training. The data preparation can be time-expensive and requires specialized skills. Depending on the application, AI approaches should be evaluated individually and compared to conventional methods in terms of effort and resulting costs. The mentioned factors can be currently limiting factors for a widespread exploitation of AI methods in industrial control and automation solutions, particularly in small to medium companies (SME). To assess the suitability of AI algorithms, Bettoni et al. [108] proposed a conceptual framework for SMEs. However, ongoing research and developments in this field are constantly showcasing new applications and pushing the technology towards industrial applications. These developments foster the wide spread of AI-based methods and facilitate the accessibility for companies. For data scare applications, transfer learning or physics-informed neural networks are promising methods to deal with this limitation and will potentially increase the interpretability of the models. Regarding control applications, reinforcement learning approaches are constantly gaining an interest in research, and new approaches and use cases will permit to understand and assess their suitability for industrial use. The combination of advanced monitoring, modelling, control and scheduling methods will in the future allow for the development of self-optimizing machines, leading to production machine improvements in terms of product quality, productivity and resource efficiency, and representing a crucial point for the next generation of human-centric manufacturing systems.

Author Contributions: Conceptualization, E.C. and D.K.; methodology, E.C. and D.K.; investigation, E.C. and D.K.; writing—original draft preparation, D.K.; writing—review and editing, E.C.; visualization, D.K.; supervision, E.C.; project administration, E.C. All authors have read and agreed to the published version of the manuscript.

Funding: This research received no external funding.

Institutional Review Board Statement: Not applicable.

Data Availability Statement: Not applicable.

Conflicts of Interest: The authors declare no conflict of interest.

References

1. Krüger, J.; Wang, L.; Verl, A.; Bauernhansl, T.; Carpanzano, E.; Makris, S.; Fleischer, J.; Reinhart, G.; Franke, J.; Pellegrinelli, S. Innovative control of assembly systems and lines. *CIRP Ann.* **2017**, *66*, 707–730. [CrossRef]
2. Jamwal, A.; Agrawal, R.; Sharma, M.; Giallanza, A. Industry 4.0 technologies for manufacturing sustainability: A systematic review and future research directions. *Appl. Sci.* **2021**, *11*, 5725. [CrossRef]
3. Chui, M.; Francisco, S.; Manyika, J. Notes from the Ai Frontier Insights from Hundreds of Use Cases. n.d. Available online: www.mckinsey.com/mgi (accessed on 10 September 2022).
4. Weichert, D.; Link, P.; Stoll, A.; Rüping, S.; Ihlenfeldt, S.; Wrobel, S. A review of machine learning for the optimization of production processes. *Int. J. Adv. Manuf. Technol.* **2019**, *104*, 1889–1902. [CrossRef]
5. Yang, L.; Zhu, M. Review on the status and development trend of AI industry. In Proceedings of the 4th International Conference on Cloud Computing and Big Data Analytics (ICCCBDA 2019), Chengdu, China, 12–15 April 2019; pp. 89–93. [CrossRef]
6. Kim, D.H.; Kim, T.J.Y.; Wang, X.; Kim, M.; Quan, Y.J.; Oh, J.W.; Min, S.H.; Kim, H.; Bhandari, B.; Yang, I.; et al. Smart Machining Process Using Machine Learning: A Review and Perspective on Machining Industry. *Int. J. Precis. Eng. Manuf. Green Technol.* **2018**, *5*, 555–568. [CrossRef]
7. Lee, C.; Lim, C. From technological development to social advance: A review of Industry 4.0 through machine learning. *Technol. Forecast. Soc. Chang.* **2021**, *167*, 120653. [CrossRef]
8. Elmenreich, W. *An Introduction to Sensor Fusion*; Vienna University of Technology: Vienna, Austria, 2002; pp. 1–28.
9. White, F.E. Joint Directors of Laboratories, Technical Panel for C3. In *Data Fusion Subpanel*; Naval Ocean Systems Center: San Diego, CA, USA, 1991.
10. Ayed, S.B.; Trichili, H.; Alimi, A.M. Data fusion architectures: A survey and comparison. In Proceedings of the 2015 15th International Conference on Intelligent Systems Design and Applications (ISDA), Marrakesh, Morocco, 14–16 December 2015; pp. 277–282. [CrossRef]
11. Azimirad, E.; Haddadnia, J.; Izadipour, A. A comprehensive review of the multi-sensor data fusion architectures. *J. Appl. Inf. Technol.* **2015**, *71*, 33–42.
12. Steinberg, A.N.; Bowman, C.L.; White, F.E. Revisions to the JDL data fusion model. In Proceedings of the AeroSense '99, Sensor Fusion: Architectures, Algorithms, and Applications III, Orlando, FL, USA, 12 March 1999; Volume 3719, p. 430. [CrossRef]
13. Meng, T.; Jing, X.; Yan, Z.; Pedrycz, W. A survey on machine learning for data fusion. *Inf. Fusion* **2020**, *57*, 115–129. [CrossRef]
14. Sasiadek, J.Z. Sensor fusion. *Annu. Rev. Control* **2002**, *26*, 203–228. [CrossRef]
15. Julier, S.J.; Uhlmann, J.K. New extension of the Kalman filter to nonlinear systems. In Proceedings of the AeroSense '97, Signal Processing. Sensor Fusion, and Target Recognition VI, Orlando, FL, USA, 28 July 1997; Volume 3068, p. 182. [CrossRef]
16. Crisan, D.; Doucet, A. A survey of convergence results on particle filtering methods for practitioners. *IEEE Trans. Signal Processing* **2002**, *50*, 736–746. [CrossRef]
17. Challa, S.; Palaniswami, M.; Shilton, A. Distributed data fusion using support vector machines. In Proceedings of the 5th International Conference on Information Fusion (FUSION 2002), Annapolis, MD, USA, 8–11 July 2002; Volume 2, pp. 881–885. [CrossRef]
18. Fung, M.L.; Chen, M.Z.Q.; Chen, Y.H. Sensor fusion: A review of methods and applications. In Proceedings of the 29th Chinese Control and Decision Conference (CCDC), Chongqing, China, 28–30 May 2017; pp. 3853–3860. [CrossRef]
19. Goebel, K.; Agogino, A.M. Fuzzy sensor fusion for gas turbine power plants. In Proceedings of the AeroSense '99, Sensor Fusion: Architectures, Algorithms, and Applications III, Orlando, FL, USA, 12 March 1999; Volume 3719, p. 52. [CrossRef]
20. Yazdkhasti, S.; Sasiadek, J.Z. Multi Sensor Fusion Based on Adaptive Kalman Filtering. In *Advances in Aerospace Guidance, Navigation and Control*; Dołęga, B., Głębocki, R., Kordos, D., Żugaj, M., Eds.; Springer: Cham, Switzerland, 2018. [CrossRef]
21. Assa, A.; Janabi-Sharifi, F. A Kalman filter-based framework for enhanced sensor fusion. *IEEE Sens J.* **2015**, *15*, 3281–3292. [CrossRef]
22. Rigatos, G.G. Extended Kalman and Particle Filtering for sensor fusion in motion control of mobile robots. *Math. Comput. Simul.* **2010**, *81*, 590–607. [CrossRef]
23. Pham, M.; Yang, D.; Sheng, W. A Sensor Fusion Approach to Indoor Human Localization Based on Environmental and Wearable Sensors. *IEEE Trans. Autom. Sci. Eng.* **2019**, *16*, 339–350. [CrossRef]
24. Bezerra, E.D.C.; Teles, A.S.; Coutinho, L.R.; Silva, F.J.D.E. Dempster–shafer theory for modeling and treating uncertainty in iot applications based on complex event processing. *Sensors* **2021**, *21*, 1863. [CrossRef] [PubMed]
25. Koksalmis, E.; Kabak, Ö. Sensor fusion based on Dempster-Shafer theory of evidence using a large scale group decision making approach. *Int. J. Intell. Syst.* **2020**, *35*, 1126–1162. [CrossRef]
26. Vitola, J.; Pozo, F.; Tibaduiza, D.A.; Anaya, M. A sensor data fusion system based on k-nearest neighbor pattern classification for structural health monitoring applications. *Sensors* **2017**, *17*, 417. [CrossRef]
27. Nweke, H.F.; Teh, Y.W.; Mujtaba, G.; Alo, U.R.; Al-garadi, M.A. Multi-sensor fusion based on multiple classifier systems for human activity identification. *Hum. -Cent. Comput. Inf. Sci.* **2019**, *9*, 34. [CrossRef]
28. Zhang, J.-W.; Sun, L.-P.; Cao, J. SVM for Sensor Fusion-a Comparison with Multilayer Perceptron Networks. In Proceedings of the 2006 International Conference on Machine Learning and Cybernetics, Dalian, China, 13–16 August 2006; pp. 2979–2984. [CrossRef]

29. Farias, G.; Fabregas, E.; Peralta, E.; Vargas, H.; Hermosilla, G.; Garcia, G.; Dormido, S. A neural network approach for building an obstacle detection model by fusion of proximity sensors data. *Sensors* **2018**, *18*, 683. [CrossRef]
30. Hu, G. Neural network applications in sensor fusion for a mobile robot motion. In Proceedings of the WASE International Conference on Information Engineering (ICIE 2010), Beidai, China, 14–15 August 2010; pp. 46–49. [CrossRef]
31. Petrich, J.; Snow, Z.; Corbin, D.; Reutzel, E.W. Multi-modal sensor fusion with machine learning for data-driven process monitoring for additive manufacturing. *Addit. Manuf.* **2021**, *48*, 102364. [CrossRef]
32. Pfeifer, T.; Protzel, P. Robust Sensor Fusion with Self-Tuning Mixture Models. In Proceedings of the IEEE International Conference on Intelligent Robots and Systems, Madrid, Spain, 1–5 October 2018; pp. 3678–3685. [CrossRef]
33. Vasic, M.; Martinoli, A. A Collaborative Sensor Fusion Algorithm for Multi-object Tracking Using a Gaussian Mixture Probability Hypothesis Density Filter. In Proceedings of the IEEE Conference on Intelligent Transportation Systems (ITSC), Gran Canaria, Spain, 15–18 September 2015; pp. 491–498. [CrossRef]
34. Stief, A.; Ottewill, J.R.; Baranowski, J.; Orkisz, M. A PCA and Two-Stage Bayesian Sensor Fusion Approach for Diagnosing Electrical and Mechanical Faults in Induction Motors. *IEEE Trans. Ind. Electron.* **2019**, *66*, 9510–9520. [CrossRef]
35. Mehta, P.; Werner, A.; Mears, L. Condition based maintenance-systems integration and intelligence using Bayesian classification and sensor fusion. *J. Intell. Manuf.* **2015**, *26*, 331–346. [CrossRef]
36. Chen, S.; Deng, Y.; Wu, J. Fuzzy sensor fusion based on evidence theory and its application. *Appl. Artif. Intell.* **2013**, *27*, 235–248. [CrossRef]
37. Shitsukane, A.; Cheruiyot, W.; Otieno, C.; Mvurya, M. Fuzzy logic sensor fusion for obstacle avoidance mobile robot. In Proceedings of the 2018 IST-Africa Week Conference IST-Africa, Gaborone, Botswana, 9–11 May 2018; pp. 108–112.
38. Liggins, M.E.; Hall, D.L.; Llinas, J. *Handbook of Multisensor Data Fusion: Theory and Practice*; CRC Press: Boca Raton, FL, USA, 2009.
39. King, R.C.; Villeneuve, E.; White, R.J.; Sherratt, R.S.; Holderbaum, W.; Harwin, W.S. Application of data fusion techniques and technologies for wearable health monitoring. *Med. Eng. Phys.* **2017**, *42*, 1–12. [CrossRef] [PubMed]
40. Li, W.; Wang, Z.; Wei, G.; Ma, L.; Hu, J.; Ding, D. A Survey on Multisensor Fusion and Consensus Filtering for Sensor Networks. *Discrete Dyn. Nat. Soc.* **2015**, *2015*, 683701. [CrossRef]
41. Olfati-Saber, R.; Shamma, J.S. Consensus filters for sensor networks and distributed sensor fusion. In Proceedings of the 44th IEEE Conference on Decision and Control, and the European Control Conference (CDC-ECC '05), Seville, Spain, 15 December 2005; Volume 2005, pp. 6698–6703. [CrossRef]
42. Wang, Z.; Wu, Y.; Niu, Q. Multi-Sensor Fusion in Automated Driving: A Survey. *IEEE Access* **2020**, *8*, 2847–2868. [CrossRef]
43. Yeong, D.J.; Velasco-Hernandez, G.; Barry, J.; Walsh, J. Sensor and Sensor Fusion Technology in Autonomous Vehicles: A Review Autonomous Systems, Data Analytics, Remote Sensing View project PROPAT Integrated Process Control View project Sensor and Sensor Fusion Technology in Autonomous Vehicles: A Review. *Sensors* **2021**, *21*, 2140. [CrossRef]
44. Alatise, M.B.; Hancke, G.P. A Review on Challenges of Autonomous Mobile Robot and Sensor Fusion Methods. *IEEE Access* **2020**, *8*, 39830–39846. [CrossRef]
45. Wei, P.; Cagle, L.; Reza, T.; Ball, J.; Gafford, J. LiDAR and camera detection fusion in a real-time industrial multi-sensor collision avoidance system. *Electronics* **2018**, *7*, 84. [CrossRef]
46. Mourtzis, D.; Doukas, M.; Bernidaki, D. Simulation in manufacturing: Review and challenges. *Procedia CIRP* **2014**, *25*, 213–229. [CrossRef]
47. Bikas, H.; Stavropoulos, P.; Chryssolouris, G. Additive manufacturing methods and modeling approaches: A critical review. *Int. J. Adv. Manuf. Technol.* **2016**, *83*, 389–405. [CrossRef]
48. Lecun, Y.; Bengio, Y.; Hinton, G. Deep learning. *Nature* **2015**, *521*, 436–444. [CrossRef]
49. Negahban, A.; Smith, J.S. Simulation for manufacturing system design and operation: Literature review and analysis. *J. Manuf. Syst.* **2014**, *33*, 241–261. [CrossRef]
50. Knüttel, D.; Baraldo, S.; Valente, A.; Bleicher, F.; Wegener, K.; Carpanzano, E. Machine learning based track height prediction for complex tool paths in direct metal deposition. *CIRP Ann.* **2022**, *71*, 193–196. [CrossRef]
51. Caiazzo, F.; Caggiano, A. Laser direct metal deposition of 2024 al alloy: Trace geometry prediction via machine learning. *Materials* **2018**, *11*, 444. [CrossRef] [PubMed]
52. Kara, F.; Aslantas, K.; Çiçek, A. ANN and multiple regression method-based modelling of cutting forces in orthogonal machining of AISI 316L stainless steel. *Neural Comput. Appl.* **2015**, *26*, 237–250. [CrossRef]
53. Kara, F.; Karabatak, M.; Ayyildiz, M.; Nas, E. Effect of machinability, microstructure and hardness of deep cryogenic treatment in hard turning of AISI D2 steel with ceramic cutting. *J. Mater. Res. Technol.* **2020**, *9*, 969–983. [CrossRef]
54. Qi, X.; Chen, G.; Li, Y.; Cheng, X.; Li, C. Applying Neural-Network-Based Machine Learning to Additive Manufacturing: Current Applications, Challenges, and Future Perspectives. *Engineering* **2019**, *5*, 721–729. [CrossRef]
55. Knüttel, D.; Baraldo, S.; Valente, A.; Wegener, K.; Carpanzano, E. Transfer learning of neural network based process models in Direct Metal Deposition. *Procedia CIRP* **2022**, *107*, 863–868. [CrossRef]
56. Knüttel, D.; Baraldo, S.; Valente, A.; Wegener, K.; Carpanzano, E. Model based learning for efficient modelling of heat transfer dynamics. *Procedia CIRP* **2021**, *102*, 252–257. [CrossRef]
57. Raissi, M.; Perdikaris, P.; Karniadakis, G.E. Physics-informed neural networks: A deep learning framework for solving forward and inverse problems involving nonlinear partial differential equations. *J. Comput. Phys.* **2019**, *378*, 686–707. [CrossRef]

58. Raissi, M.; Karniadakis, G.E. Hidden physics models: Machine learning of nonlinear partial differential equations. *J. Comput. Phys.* **2018**, *357*, 125–141. [CrossRef]
59. N Zobeiry, K.D.H. A Physics-Informed Machine Learning Approach for Solving Heat Transfer Equation in Advanced Manufacturing and Engineering Applications. *arXix* **2020**, arXiv:2010.02011. [CrossRef]
60. Li, Z.; Kovachki, N.; Azizzadenesheli, K.; Liu, B.; Bhattacharya, K.; Stuart, A.; Anandkumar, A. Fourier Neural Operator for Parametric Partial Differential Equations. *arXiv* **2020**, arXiv:2010.08895.
61. Zhang, K.; Zuo, Y.; Zhao, H.; Ma, X.; Gu, J.; Wang, J.; Yang, Y.; Yao, C.; Yao, J. Fourier Neural Operator for Solving Subsurface Oil/Water Two-Phase Flow Partial Differential Equation. *SPE J.* **2022**, *27*, 1815–1830. [CrossRef]
62. Maier, M.; Rupenyan, A.; Bobst, C.; Wegener, K. Self-optimizing grinding machines using Gaussian process models and constrained Bayesian optimization. *Int. J. Adv. Manuf. Technol.* **2020**, *108*, 539–552. [CrossRef]
63. Shin, J.; Badgwell, T.A.; Liu, K.; Lee, J.H. Reinforcement Learning—Overview of recent progress and implications for process control. *Comput. Chem. Eng.* **2019**, *127*, 282–294. [CrossRef]
64. Wuest, T.; Weimer, D.; Irgens, C.; Thoben, K.D. Machine learning in manufacturing: Advantages, challenges, and applications. *Prod. Manuf. Res.* **2016**, *4*, 23–45. [CrossRef]
65. Mousavi, S.S.; Schukat, M.; Howley, E. Deep Reinforcement Learning: An Overview. In *Proceedings of SAI Intelligent Systems Conference*; Lecture Notes in Networks and Systems; Springer: Cham, Switzerland, 2018; Volume 16, pp. 426–440. [CrossRef]
66. Nian, R.; Liu, J.; Huang, B. A review on reinforcement learning: Introduction and applications in industrial process control. *Comput. Chem. Eng.* **2020**, *139*, 106886. [CrossRef]
67. Ziegler, J.G.; Nichols, N. Optimum Settings for Automatic Controllers. *J. Dyn. Syst. Meas. Control. Trans. ASME* **1942**, *115*, 220–222. [CrossRef]
68. Wang, X.-S.; Cheng, Y.-H.; Sun, W. A Proposal of Adaptive PID Controller Based on Reinforcement Learning. *J. China Univ. Min. Technol.* **2007**, *17*, 40–44. [CrossRef]
69. Lee, J.H.; Shin, J.; Realff, M.J. Machine learning: Overview of the recent progresses and implications for the process systems engineering field. *Comput. Chem. Eng.* **2018**, *114*, 111–121. [CrossRef]
70. Lin, Y.; McPhee, J.; Azad, N.L. Comparison of Deep Reinforcement Learning and Model Predictive Control for Adaptive Cruise Control. *IEEE Trans. Intell. Veh.* **2021**, *6*, 221–231. [CrossRef]
71. Shah, P.; Agashe, S. Review of fractional PID controller. *Mechatronics* **2016**, *38*, 29–41. [CrossRef]
72. Holcomb, S.D.; Porter, W.K.; Ault, S.V.; Mao, G.; Wang, J. Overview on DeepMind and its AlphaGo Zero AI. In Proceedings of the International Conference on Big Data and Education, Honolulu HI USA, 9 March 2018; pp. 67–71. [CrossRef]
73. Garcia, J.; Fernández, F. A Comprehensive Survey on Safe Reinforcement Learning. *J. Mach. Learn. Res.* **2015**, *16*, 1437–1480.
74. Brunke, L.; Greeff, M.; Hall, A.W.; Yuan, Z.; Zhou, S.; Panerati, J.; Schoellig, A.P. Safe Learning in Robotics: From Learning-Based Control to Safe Reinforcement Learning. *Annu. Rev. Control Robot. Auton. Syst.* **2022**, *5*, 411–444. [CrossRef]
75. Cardin, O.; Trentesaux, D.; Thomas, A.; Castagna, P.; Berger, T.; El-Haouzi, H.B. Coupling predictive scheduling and reactive control in manufacturing hybrid control architectures: State of the art and future challenges. *J. Intell. Manuf.* **2017**, *28*, 1503–1517. [CrossRef]
76. Schuh, G.; Reuter, C.; Prote, J.P.; Brambring, F.; Ays, J. Increasing data integrity for improving decision making in production planning and control. *CIRP Ann. Manuf. Technol.* **2017**, *66*, 425–428. [CrossRef]
77. Mourtzis, D.; Angelopoulos, J.; Panopoulos, N. Robust engineering for the design of resilient manufacturing systems. *Appl. Sci.* **2021**, *11*, 3067. [CrossRef]
78. Carpanzano, E.; Jovane, F. Advanced Automation Solutions for Future Adaptive Factories. *CIRP Ann. Manuf. Technol.* **2007**, *56*, 435–438. [CrossRef]
79. Ecarpanzano; Cesta, A.; Orlandini, A.; Rasconi, R.A. Valente, Intelligent dynamic part routing policies in Plug&Produce Reconfigurable Transportation Systems. *CIRP Ann. Manuf. Technol.* **2014**, *63*, 425–428. [CrossRef]
80. Valente, A.; Carpanzano, E. Development of multi-level adaptive control and scheduling solutions for shop-floor automation in reconfigurable manufacturing systems. *CIRP Ann. Manuf. Technol.* **2011**, *60*, 449–452. [CrossRef]
81. Cadavid, J.P.U.; Lamouri, S.; Grabot, B.; Pellerin, R.; Fortin, A. Machine learning applied in production planning and control: A state-of-the-art in the era of industry 4. *0. J. Intell. Manuf.* **2020**, *31*, 1531–1558. [CrossRef]
82. Montazeri, M.; van Wassenhove, L.N. Analysis of scheduling rules for an FMS. *Int. J. Prod. Res.* **1990**, *28*, 785–802. [CrossRef]
83. Pach, C.; Berger, T.; Bonte, T.; Trentesaux, D. ORCA-FMS: A dynamic architecture for the optimized and reactive control of flexible manufacturing scheduling. *Comput. Ind.* **2014**, *65*, 706–720. [CrossRef]
84. Zhang, J.; Ding, G.; Zou, Y.; Qin, S.; Fu, J. Review of job shop scheduling research and its new perspectives under Industry 4.0. *J. Intell. Manuf.* **2019**, *30*, 1809–1830. [CrossRef]
85. Ouelhadj, D.; Petrovic, S. A survey of dynamic scheduling in manufacturing systems. *J. Sched.* **2009**, *12*, 417–431. [CrossRef]
86. Barbosa, J.; Leitão, P.; Adam, E.; Trentesaux, D. Nervousness in dynamic self-organized holonic multi-agent systems, Advances in Intelligent and Soft Computing. In *Highlights on Practical Applications of Agents and Multi-Agent Systems*; Springer: Berlin/Heidelberg, Germany, 2012; Volume 156, pp. 9–17. [CrossRef]
87. Carvalho, T.P.; Soares, F.A.A.M.N.; Vita, R.; Francisco, R.D.; Basto, J.P.; Alcalá, S.G.S. A systematic literature review of machine learning methods applied to predictive maintenance. *Comput. Ind. Eng.* **2019**, *137*, 106024. [CrossRef]

88. Selcuk, S. Predictive maintenance, its implementation and latest trends. *Proc. Inst. Mech. Eng. B J. Eng. Manuf.* **2017**, *231*, 1670–1679. [CrossRef]
89. Priore, P.; Ponte, B.; Puente, J.; Gómez, A. Learning-based scheduling of flexible manufacturing systems using ensemble methods. *Comput. Ind. Eng.* **2018**, *126*, 282–291. [CrossRef]
90. Lubosch, M.; Kunath, M.; Winkler, H. Industrial scheduling with Monte Carlo tree search and machine learning. *Procedia CIRP* **2018**, *72*, 1283–1287. [CrossRef]
91. De Puiseau, C.W.; Meyes, R.; Meisen, T. On reliability of reinforcement learning based production scheduling systems: A comparative survey. *J. Intell. Manuf.* **2022**, *33*, 911–927. [CrossRef]
92. D'Addona, D.M.; Bracco, F.; Bettoni, A.; Nishino, N.; Carpanzano, E.; Bruzzone, A.A. Adaptive automation and human factors in manufacturing: An experimental assessment for a cognitive approach. *CIRP Ann.* **2018**, *67*, 455–458. [CrossRef]
93. Wang, L.; Gao, R.; Váncza, J.; Krüger, J.; Wang, X.V.; Makris, S.; Chryssolouris, G. Symbiotic human-robot collaborative assembly. *CIRP Ann.* **2019**, *68*, 701–726. [CrossRef]
94. Kemény, Z.; Váncza, J.; Wang, L.; Wang, X.V. Human–Robot Collaboration in Manufacturing: A Multi-agent View. In *Advanced Human-Robot Collaboration in Manufacturing*; Wang, L., Wang, X.V., Váncza, J., Kemény, Z., Eds.; Springer International Publishing: Cham, Switzerland, 2021; pp. 3–41. [CrossRef]
95. Bettoni, A.; Montini, E.; Righi, M.; Villani, V.; Tsvetanov, R.; Borgia, S.; Secchi, C.; Carpanzano, E. Mutualistic and adaptive human-machine collaboration based on machine learning in an injection moulding manufacturing line. *Procedia CIRP* **2020**, *93*, 395–400. [CrossRef]
96. Valente, A.; Pavesi, G.; Zamboni, M.; Carpanzano, E. Deliberative robotics—A novel interactive control framework enhancing human-robot collaboration. *CIRP Ann.* **2022**, *71*, 21–24. [CrossRef]
97. Bracco, F.; Bruzzone, A.A.; Carpanzano, E. Transfactory: Towards a New Technology-Human Manufacturing Co-evolution Framework. In *Advances in System-Integrated Intelligence*; Valle, M.; Lehmhus, D., Gianoglio, C., Ragusa, E., Seminara, L., Bosse, S., Ibrahim, A., Thoben, K.-D., Eds.; Springer International Publishing: Cham, Switzerland, 2023; pp. 636–645.
98. Klocke, F.; Abel, D.; Hopmann, C.; Auerbach, T.; Keitzel, G.; Reiter, M.; Reßmann, A.; Stemmler, S.; Veselovac, D. Approaches of Self-optimising Systems in Manufacturing. In *Advances in Production Technology*; Springer: Cham, Switzerland, 2015; pp. 161–173. [CrossRef]
99. Brettel, M.; Fischer, F.G.; Bendig, D.; Weber, A.R.; Wolff, B. Enablers for Self-optimizing Production Systems in the Context of Industrie 4.0. *Procedia CIRP* **2016**, *41*, 93–98. [CrossRef]
100. Simeone, A.; Zeng, Y.; Caggiano, A. Intelligent decision-making support system for manufacturing solution recommendation in a cloud framework. *Int. J. Adv. Manuf. Technol.* **2021**, *112*, 1035–1050. [CrossRef]
101. Schmitt, R.; Brecher, C.; Corves, B.; Gries, T.; Jeschke, S.; Klocke, F.; Loosen, P.; Michaeli, W.; Müller, R.; Poprawe, R.; et al. *Integrative Production Technology for High-Wage Countries*; Springer: Berlin/Heidelberg, Germany, 2012. [CrossRef]
102. Permin, E.; Bertelsmeier, F.; Blum, M.; Bützler, J.; Haag, S.; Kuz, S.; Özdemir, D.; Stemmler, S.; Thombansen, U.; Schmitt, R.; et al. Self-optimizing production systems. *Procedia CIRP* **2016**, *41*, 417–422. [CrossRef]
103. Thombansen, U.; Buchholz, G.; Frank, D.; Heinisch, J.; Kemper, M.; Pullen, T.; Reimer, V.; Rotshteyn, G.; Schwenzer, M.; Stemmler, S.; et al. Design framework for model-based self-optimizing manufacturing systems. *Int. J. Adv. Manuf. Technol.* **2018**, *97*, 519–528. [CrossRef]
104. Möhring, H.C.; Wiederkehr, P.; Erkorkmaz, K.; Kakinuma, Y. Self-optimizing machining systems. *CIRP Ann.* **2020**, *69*, 740–763. [CrossRef]
105. Monostori, L. AI and machine learning techniques for managing complexity, changes and uncertainties in manufacturing. *Eng. Appl. Artif. Intell.* **2003**, *16*, 277–291. [CrossRef]
106. Mayr, J.; Blaser, P.; Ryser, A.; Hernandez-Becerro, P. An adaptive self-learning compensation approach for thermal errors on 5-axis machine tools handling an arbitrary set of sample rates. *CIRP Ann.* **2018**, *67*, 551–554. [CrossRef]
107. Dittrich, M.A.; Uhlich, F.; Denkena, B. Self-optimizing tool path generation for 5-axis machining processes. *CIRP J. Manuf. Sci. Technol.* **2019**, *24*, 49–54. [CrossRef]
108. Bettoni, A.; Matteri, D.; Montini, E.; Gladysz, B.; Carpanzano, E. An AI adoption model for SMEs: A conceptual framework. *IFAC-Pap.* **2021**, *54*, 702–708. [CrossRef]

applied sciences

MDPI

Article

Adaptive Optimal Robust Control for Uncertain Nonlinear Systems Using Neural Network Approximation in Policy Iteration

Dengguo Xu [1,2], Qinglin Wang [1] and Yuan Li [1,*]

1 School of Automation, Beijing Institute of Technology, Beijing 100081, China; xudengguo@bit.edu.cn (D.X.); wangql@bit.edu.cn (Q.W.)
2 School of Electrical Engineering, Liupanshui Normal University, Liupanshui 553004, China
* Correspondence: liyuan@bit.edu.cn

Abstract: In this study, based on the policy iteration (PI) in reinforcement learning (RL), an optimal adaptive control approach is established to solve robust control problems of nonlinear systems with internal and input uncertainties. First, the robust control is converted into solving an optimal control containing a nominal or auxiliary system with a predefined performance index. It is demonstrated that the optimal control law enables the considered system globally asymptotically stable for all admissible uncertainties. Second, based on the Bellman optimality principle, the online PI algorithms are proposed to calculate robust controllers for the matched and the mismatched uncertain systems. The approximate structure of the robust control law is obtained by approximating the optimal cost function with neural network in PI algorithms. Finally, in order to illustrate the availability of the proposed algorithm and theoretical results, some numerical examples are provided.

Keywords: policy iteration; uncertain nonlinear system; robust control; adaptive optimal control

Citation: Xu, D.; Wang, Q.; Li, Y. Adaptive Optimal Robust Control for Uncertain Nonlinear Systems Using Neural Network Approximation in Policy Iteration. *Appl. Sci.* **2021**, *11*, 2312. https://doi.org/10.3390/app11052312

Academic Editor: Emanuele Carpanzano and Giovanni Petrone

Received: 21 January 2021
Accepted: 28 February 2021
Published: 5 March 2021

1. Introduction

It is ineluctable to contain uncertain parameters and disturbances in practical systems due to modeling errors, external disturbances, and so on [1]. Thus, it is of great practical significance to study robust control of the uncertain systems. In recent years, the control problem of uncertain systems has been extensively studied. The literature on robust control of uncertain systems mainly includes linear systems and nonlinear systems. For the uncertain linear systems, algebraic Riccati equations (ARE) or linear matrix inequalities (LMI) were mostly used to deal with them in the classical research methods [2–6]. In this literature, both matched and mismatched systems were involved. For nonlinear systems, the early research methods include feedback linearization, fuzzy modeling, nonlinear H_∞ control, and so on [7–11]. However, in recent decades, neural networks (NN) and PI in RL were used to approximate the robust control law numerically [12–14].

The PI method in RL was initially utilized to calculate the optimal control law for some deterministic systems. Werbos first proposed an idea of approximating the solution in Bellman's equation using approximate dynamic programming (ADP) [15]. There have been many results of using the PI method to calculate the optimal control law for some deterministic systems [16–19]. There are two major benefits of using the PI algorithm to deal with such optimal problems. On the one hand, it can effectively solve some problems of the "curse of dimensionality" in engineering control [20]. On the other hand, it can be utilized to calculate the optimal control law without knowing the system dynamics. In the practice of engineering control, it is difficult to obtain system dynamics accurately. Therefore, it is a good choice to use the PI algorithm to solve unknown model control problems.

Within the last ten years, the PI method was also developed to calculate the robust controller for some uncertain systems, which is based on the optimal control method of robust control [21]. For an input constraint nonlinear system with continuous time, a novel RL-based algorithm was prosposed to deal with the robust control problem in [22]. Based on network structure approximation, an online PI algorithm was developed to solve robust control of a class of nonlinear discrete-time uncertain systems in [23]. Using a data-driven RL algorithm, a robust control scheme was developed for a class of completely unknown dynamic systems with uncertainty in [24]. In addition, there are many other examples of literature on robust control based on RL, such as [25–27]. In all the literature listed above, the solution to the Hamilton Jacobi Bellman (HJB) equation was approximated by neural network. In fact, solving the HJB equation is a key problem in optimal control problem [28]. The HJB equation is difficult to solve because it is a nonlinear partial differential equation. For a nonlinear system, the HJB equation is solved with neural network approximation in many cases. Meanwhile, for the linear system, ARE is used to solve it instead of a neural network. However, for all we know, most of the current studies have not considered the input uncertainty in system. The input uncertainty does exist in the actual control system.

In this study, a class of continuous-time nonlinear systems with internal and input uncertainties is considered. The main objective is to establish robust control laws for the uncertain systems. By solving the optimal control problem constructed, the robust control problem is converted into calculating an optimal controller. The online PI algorithms are proposed to calculate robust control by approximating the optimal cost with neural network. The convergence of the proposed algorithms is proved. Numerical examples are given to illustrate the availability of the method.

Our main contributions are as follows. First, more general uncertain nonlinear systems are considered, in which the uncertainty entered both the system and the input. For the matched and the mismatched uncertain systems, it is proved that the robust control can be converted into calculating an optimal controller. Second, the online PI algorithms are developed to solve the robust control problem. The neural network is utilized to approximate the optimal cost in PI algorithm, which fulfilled a difficult task of solving the HJB equation.

The rest of this paper is arranged as follows. We formulate the robust control problems and propose some basic results for the issues under consideration in Section 2. Solving the robust control problem is converted to calculate an optimal control law of a nominal or auxiliary system in Sections 3 and 4. Based on approximating optimal cost with neural network, the online PI algorithms are developed to solve the robust control problem in Section 5. To support the proposed theoretical framework, we provide some numerical examples in Section 6. In Section 7, the study is concluded, and the scope for future research is discussed.

2. Preliminaries and Problem Formulation

Consider an uncertain nonlinear system as follows:

$$\dot{x}(t) = f(x(t)) + \Delta f(x(t)) + [g(x(t)) + \Delta g(x(t))]u(t), x(0) = x_0 \tag{1}$$

where $x(t) \in \mathbb{R}^n$ is the system state, $u(t) \in \mathbb{R}^m$ is the control input, $f(x(t)) \in \mathbb{R}^n$, $g(x(t)) \in \mathbb{R}^{n\times m}$ are known function, $\Delta f(x(t)) \in \mathbb{R}^n$, $\Delta g(x(t)) \in \mathbb{R}^{n\times m}$ are uncertain disturbing function.

The control objective is to establish a robust control law $u = u(x)$ in order that the closed-loop system is asymptotically stable for all allowed uncertain disturbances $\Delta f(x(t))$ and $\Delta g(x(t))$.

As a general case, we first make the following assumptions to ensure that the state Equation (1) is well defined [1,29].

Assumption 1. *In (1), $f(x) + g(x)u$ is Lipschitz continuous with respect to x and u on the set $\Omega \subseteq \mathbb{R}^n$ containing the origin.*

Assumption 2. *For the free vibration system, $f(0) = 0, \Delta f(0) = 0$, that is, $x = 0$ is the equilibrium point of the free vibration system.*

Definition 1. *The system (1) is called to satisfy the system dynamics matched condition if there is a function matrix $h(x) \in R^{m\times 1}$ such that*

$$\Delta f(x) = g(x)h(x). \tag{2}$$

Definition 2. *System (1) is called to satisfy the input matched condition if there is a function $m(x) \in R^{m\times m}$ such that*

$$\Delta g(x) = g(x)m(x), \tag{3}$$

where $m(x) \geq 0$.

Definition 3. *If the system (1) satisfies the conditions (2) and (3) for any allowed disturbances $\Delta f(x)$ and $\Delta g(x)$, then the system (1) is called a matched uncertain system.*

Definition 4. *If the system (1) does not satisfy the condition (2) or (3) for any allowed disturbances $\Delta f(x)$ and $\Delta g(x)$, then the system (1) is called a mismatched uncertain system.*

Next, we consider the robust control problem of nonlinear system (1) with matched and mismatched conditions, respectively.

3. Robust Control of Matched Uncertain Nonlinear Systems

This section considers the problem of robust control when the system (1) meets the matched conditions (2) and (3). By constructing appropriate performance indexes, the problem of robust control is transformed into calculating the optimal control law of a corresponding nominal system. Based on the optimal control of the nominal system, a PI algorithm is proposed to obtain robust feedback controller.

For the nominal system

$$\dot{x} = f(x) + g(x)u, \tag{4}$$

find the controller $u = u(x)$ to minimize performance index

$$J(x_0, u) = \int_0^\infty [f_{max}^2(x) + x^T x + u^T u]dt \tag{5}$$

where $f_{max}(x)$ is the supremum function of uncertainty $h(x)$, that is $\|h(x)\| \leq f_{max}(x)$.

The definition of admissible control in optimal control problem is given below [26].

Definition 5. *The control policy $u(x)$ is called an admissible control of the system (4) with regard to the performance function (5) on compact set $\Omega \subseteq \mathbb{R}^n$ if $u(x)$ is continuous on Ω, $u(0) = 0$, it can stabilize the system (4) on Ω, and the performance function (5) is limited for any $x \in \Omega$.*

According to the performance index (5), the cost function corresponding to the admissible control $u(x)$ is given by

$$V(x, u) = \int_t^\infty [f_{max}^2(x) + x^T x + u^T u]dt \tag{6}$$

Taking time derivative on both side of (6), it follows the Bellman equation

$$f_{max}^2(x) + x^T x + u^T u + \nabla V^T [f(x) + g(x)u] = 0, \tag{7}$$

where ∇V is the gradient vector of the cost function $V(x, u)$ with respect to x.

Definite Hamiltonian function

$$H(x, u, \nabla V) = f_{max}^2(x) + x^T x + u^T u + \nabla V^T [f(x) + g(x)u] \tag{8}$$

Determining the extremum of the Hamiltonian function yields the optimal control function

$$u^*(x) = -\frac{1}{2}g^T(x)\nabla V \tag{9}$$

By substituting (9) into (7), it follows that optimal cost $V^*(x)$ satisfies the following HJB equation

$$f_{max}^2(x) + x^T x + \nabla V^{*T} f(x) - \frac{1}{4}\nabla V^{*T} g(x) g^T(x) \nabla V^* = 0 \tag{10}$$

and initial conditions $V^*(0) = 0$.

Solving the optimal cost $V^*(x)$ from the HJB Equation (10), we can get the solution to the optimal control problem. Thus, the robust control problem can be solved.

The following theorem shows that optimal control $u^*(x) = -\frac{1}{2}g^T(x)\nabla V^*$ is a robust controller for matched uncertain systems.

Theorem 1. *Assume that the conditions (2) and (3) hold in system (1) and the solution $V^*(x)$ in HJB Equation (10) exists. Considering the nominal nonlinear system (4) with performance index (5), then the optimal control policy $u^*(x) = -\frac{1}{2}g^T(x)\nabla V^*$ can globally asymptotically stabilize the nonlinear uncertain system (1). That is to say, the closed-loop uncertain system $\dot{x}(t) = f(x(t)) + \Delta f(x(t)) + [g(x(t)) + \Delta g(x(t))]u^*(x)$ is globally asymptotically stable.*

Proof. In order to prove the stability with controller $u^*(x) = -\frac{1}{2}g^T(x)\nabla V^*$, $V^*(x)$ is chosen as the Lyapunov function. Considering the performance index (5), $V^*(x)$ is obviously positive, and $V^*(0) = 0$. Taking time derivative of the function $V^*(x)$ along closed-loop system (1), it follows that

$$\frac{dV^*}{dt} = \nabla V^{*T}[f(x) + \Delta f(x)] - \frac{1}{2}\nabla V^{*T}[g(x) + \Delta g(x)]g^T(x)\nabla V^* \tag{11}$$

Using the matched conditions (2)and (3), it follows from (11) that

$$\begin{aligned}\frac{dV^*}{dt} =& \nabla V^{*T} f(x) + \nabla V^{*T} g(x) h(x) - \frac{1}{2}\nabla V^{*T} g(x) g^T(x) \nabla V^* \\ &- \frac{1}{2}\nabla V^{*T} g(x) m(x) g^T(x) \nabla V^*\end{aligned} \tag{12}$$

From HJB Equation (10), one can obtain

$$\nabla V^{*T} f(x) = -f_{max}^2(x) - x^T x + \frac{1}{4}\nabla V^{*T} g(x) g^T(x) \nabla V^* \tag{13}$$

Substituting (13) into (12) yields

$$\begin{aligned}\frac{dV^*}{dt} =& - f_{max}^2(x) - x^T x + \frac{1}{4}\nabla V^{*T} g(x) g^T(x) \nabla V^* + \nabla V^{*T} g(x) h(x) - \frac{1}{2}\nabla V^{*T} g(x) g^T(x) \nabla V^* \\ &- \frac{1}{2}\nabla V^{*T} g(x) m(x) g^T(x) \nabla V^* \\ =& - f_{max}^2(x) - x^T x - \frac{1}{4}\nabla V^{*T} g(x) g^T(x) \nabla V^* + \nabla V^{*T} g(x) h(x) - \frac{1}{2}\nabla V^{*T} g(x) m(x) g^T(x) \nabla V^*\end{aligned} \tag{14}$$

It follows from $m(x) \geq 0$ that $-\frac{1}{2}\nabla V^{*T} g(x) m(x) g^T(x) \nabla V^* \leq 0$. Therefore, from (14), we can have

$$
\begin{aligned}
\frac{dV^*}{dt} &\leq -f_{max}^2(x) - x^T x - \frac{1}{4}\nabla V^{*T} g(x) g^T(x) \nabla V^* + \nabla V^{*T} g(x) h(x) \\
&= -f_{max}^2(x) - x^T x - \frac{1}{4}[\nabla V^{*T} g(x) g^T(x) \nabla V^* - 4\nabla V^{*T} g(x) h(x) + 4h(x)^T h(x)] + h(x)^T h(x) \\
&= -x^T x + h(x)^T h(x) - f_{max}^2(x) - \frac{1}{4}[g^T(x)\nabla V^* - 2h(x)]^T [g^T(x)\nabla V^* - 2h(x)] \\
&\leq -x^T x
\end{aligned}
\tag{15}
$$

Therefore, by Lyapunov stability theory [30], the optimal control $u^*(x) = -\frac{1}{2}g^T(x)\nabla V^*$ can make the matched uncertain system (1) asymptotically stable. Thus, for a constant $c > 0$, there is a neighborhood $N = \{x : \|x\| < c\}$ such that if the state $x(t)$ enters the neighborhood N, then $x \to 0$ when $t \to \infty$. However, $x(t)$ cannot stay out of the domain N forever; otherwise, for all $t > 0$, there is $\|x(t)\| \geq c$. This implies that

$$
\begin{aligned}
V^*[x(t)] - V^*[x(0)] &= \int_0^t \dot{V}^*(x(\tau))dt \\
&\leq \int_0^t (-x^T x)dt \\
&\leq -\int_0^t c^2 dt \\
&= -c^2 t
\end{aligned}
$$

Therefore, when $t \to \infty$, $V^*[x(t)] \leq V^*[x(0)] - c^2 t \to -\infty$. This contradicts that $V^*[x(t)]$ is positive definite. Consequently, the system (1) is globally asymptotically stable. □

Remark 1. *For matched nonlinear systems, the robust controller can be obtained by solving the optimal cost function $V^*[x(t)]$ from HJB Equation (10). In Section 4, we will use the PI algorithm to solve the HJB equation, which is a difficult partial differential equation.*

4. Robust Control of Nonlinear Systems with Mismatched Uncertainties

In this section, we consider the robust control problem when the system (1) does not satisfy the matched condition (2). At this time, the system is a mismatched nonlinear uncertain system. By constructing the appropriate auxiliary system and performance index, the robust control for the mismatched uncertain system is transformed into solving optimal control law of an auxiliary system.

Firstly, the following assumptions are given.

Assumption 3. *Suppose that the uncertainty of system (1) satisfies $\Delta f(x) = c(x)h(x)$, $\Delta g(x) = g(x)m(x)$, where $c(x)$ is a known function matrix of appropriate dimensions, $h(x)$ and $m(x)$ are uncertain functions, and $m(x) \geq 0$.*

The goal of robust control is to find a control function $u(x)$, which makes the closed-loop system

$$
\dot{x} = f(x) + c(x)h(x) + [g(x) + g(x)m(x)]u(x) \tag{16}
$$

globally asymptotically stable for all uncertainties $h(x)$ and $m(x)$.

In order to obtain the robust controller, an optimal control problem is constructed as follow. For the following auxiliary systems

$$
\dot{x} = f(x) + g(x)u + [I - g(x)g(x)^+]c(x)v \tag{17}
$$

find the controller $u = u(x), v = v(x)$, such that the performance index

$$J(x_0, u) = \int_0^\infty [f_{max}^2(x) + g_{max}^2(x) + \beta^2 x^T x + u^T u + v^T v] dt \tag{18}$$

is minimized, where β is the design parameter, $g(x)^+ = [g^T(x)g(x)]^{-1}g^T(x)$ is a pseudo inverse of the matrix function $g(x)$. Moreover, $f_{max}(x)$ and $g_{max}(x)$ are nonnegative functions and satisfy the conditions

$$\|g(x)^+ c(x)h(x)\| \le f_{max}(x), \|h(x)\| \le g_{max}(x) \tag{19}$$

According to the performance index (18), the cost function corresponding to the admissible control $(u(x), v(x))$ is

$$V(x) = \int_t^\infty [f_{max}^2(x) + g_{max}^2(x) + \beta^2 x^T x + u^T u + v^T v] dt \tag{20}$$

The following Bellman equation is obtained by taking the time derivation on both sides of (20)

$$f_{max}^2(x) + g_{max}^2(x) + \beta^2 x^T x + \bar{u}^T \bar{u} + \nabla V^T [f(x) + \bar{g}(x)\bar{u}] = 0, \tag{21}$$

where ∇V is the gradient vector of $V(x)$ with respect to x, $\bar{g}(x) = [g(x), (I - g(x)g(x)^+)c(x)]$, $\bar{u} = [u^T, v^T]^T$.

Defining Hamiltonian functions as

$$H(x, u, \nabla V) = f_{max}^2(x) + g_{max}^2(x) + \beta^2 x^T x + \bar{u}^T \bar{u} + \nabla V^T [f(x) + \bar{g}(x)\bar{u}] \tag{22}$$

Assuming that the minimum value exists and is unique in (22), the optimal control law is given by

$$\begin{aligned} \bar{u}^*(x) = \begin{bmatrix} u^*(x) \\ v^*(x) \end{bmatrix} &= -\frac{1}{2}\bar{g}^T(x)\nabla V^* \\ &= -\frac{1}{2}\begin{bmatrix} g^T(x)\nabla V^* \\ c^T(x)[I - g(x)g(x)^+]^T \nabla V^* \end{bmatrix} \end{aligned} \tag{23}$$

By substituting (23) into (21), the HJB equation is given by

$$f_{max}^2(x) + g_{max}^2(x) + \beta^2 x^T x + \nabla V^{*T} f(x) - \frac{1}{4}\nabla V^{*T} \bar{g}(x)\bar{g}^T(x)\nabla V^* = 0 \tag{24}$$

and the initial value $V^*(0) = 0$.

Remark 2. *Generally, the pseudo-inverse of $g(x)$, $g(x)^+$ will exist if its columns are linearly independent when Assumptions 1 and 2 are true [31]. In practical control systems, the function, $g(x)$, is usually column full-rank. Therefore, the pseudo-inverse of the function $g(x)$ is generally satisfied. Furthermore, the pseudo-inverse $g(x)^+$ satisfies $g(x)^+ g(x) = I$. However, it does not satisfy $g(x)g(x)^+ = I$. In addition, the auxiliary system constructed above is not a nominal system, but a compensation control term $v(x)$ is added to the nominal system.*

If we can choose an appropriate parameter β, the optimal cost $V^*(x)$ can be computed from HJB Equation (24). Then, we can get the optimal control law of system (17) with performance index (18). The following theorem shows that optimal control $u^*(x) = -\frac{1}{2}g^T(x)\nabla V^*$ is a robust controller for uncertain systems.

Theorem 2. *Assume that the mismatched uncertain system (16) satisfies Assumptions 4.1, 4.2 and (19). Consider the auxiliary system (17) corresponding to the performance index (18). There* *exists a solution $V^*(x)$ in HJB Equation (24) for a selected parameter β, and for a constant β' satisfying $|\beta'| < |\beta|$, such that*

$$2v^{*T}(x)v^*(x) \leq \beta'^2 x^T x$$

Then, the optimal control policy $u^*(x) = -\frac{1}{2}g^T(x)\nabla V^*$ can globally asymptotically stabilize the nonlinear uncertain system (16). That is to say, the closed-loop uncertain system $\dot{x}(t) = f(x) + c(x)h(x) + [g(x) + g(x)m(x)]u^*(x)$ is globally asymptotically stable.

Proof. In order to prove the global asymptotic stability of the closed-loop system, $V^*(x)$ is chosen as the Lyapunov function. Considering the performance index (18), $V^*(x)$ is obviously positive, and $V^*(0) = 0$. Taking the time derivative of the function $V^*(x)$ along the system (16), we have

$$\frac{dV^*}{dt} = \nabla V^{*T}[f(x) + c(x)h(x) + g(x)u^*(x)] + \nabla V^{*T} g(x)m(x)u^*(x) \tag{25}$$

Using $u^*(x) = -\frac{1}{2}g^T(x)\nabla V^*$ yields

$$\begin{aligned}
\frac{dV^*}{dt} =& \nabla V^{*T} f(x) + \nabla V^{*T} c(x)h(x) + \nabla V^{*T} g(x)u^*(x) - \frac{1}{2}\nabla V^{*T} g(x)m(x)g^T(x)\nabla V^* \\
\leq& \nabla V^{*T} f(x) + \nabla V^{*T} c(x)h(x) + \nabla V^{*T} g(x)u^*(x) \\
=& \nabla V^{*T}[f(x) + g(x)u^*(x) + c(x)h(x)] - \nabla V^{*T}(I - g(x)g(x)^+)c(x)v^*(x) \\
&+ \nabla V^{*T}(I - g(x)g(x)^+)c(x)v^*(x) \\
=& \nabla V^{*T}[f(x) + g(x)u^*(x) + (I - g(x)g(x)^+)c(x)v^*(x)] + \nabla V^{*T} g(x)g(x)^+c(x)h(x) \\
&- \nabla V^{*T}(I - g(x)g(x)^+)c(x)v^*(x) + \nabla V^{*T}(I - g(x)g(x)^+)c(x)h(x)
\end{aligned}$$

by $u^*(x) = -\frac{1}{2}g^T(x)\nabla V^*$ and $v^*(x) = -\frac{1}{2}c^T(x)(I - g(x)g(x)^+)^T\nabla V^*$,

$$\begin{aligned}
\frac{dV^*}{dt} \leq& \nabla V^{*T} f(x) - 2u^*(x)^T u^*(x) - 2v^{*T}(x)v^*(x) + \nabla V^{*T} g(x)g(x)^+c(x)h(x) \\
&- \nabla V^{*T}(I - g(x)g(x)^+)c(x)v^*(x) + \nabla V^{*T}(I - g(x)g(x)^+)c(x)h(x)
\end{aligned}$$

It follows from (24) that

$$\nabla V^{*T} f(x) = - f_{max}^2(x) - g_{max}^2(x) - \beta^2 x^T x + u^*(x)^T u^*(x) + v^{*T}(x)v^*(x)$$

As a result,

$$\begin{aligned}
\frac{dV^*}{dt} \leq& - f_{max}^2(x) - g_{max}^2(x) - \beta^2 x^T x - u^*(x)^T u^*(x) + v^{*T}(x)v^*(x) \\
&- 2u^*(x)^T g(x)^+c(x)h(x) - 2v^{*T}(x)h(x)
\end{aligned} \tag{26}$$

On the other hand

$$-2u^*(x)^T g(x)^+c(x)h(x) \leq [g(x)^+c(x)h(x)]^T[g(x)^+c(x)h(x)] + u^*(x)^T u^*(x)$$

Therefore,

$$\begin{aligned}
-u^*(x)^T u^*(x) - 2u^*(x)^T g(x)^+c(x)h(x) \leq& [g(x)^+c(x)h(x)]^T[g(x)^+c(x)h(x)] \\
\leq& f_{max}^2(x)
\end{aligned} \tag{27}$$

It follows from the basic matrix inequality that

$$\begin{aligned} -2v^{*T}(x)h(x) &\leq v^{*T}(x)v^*(x) + h^T(x)h(x) \\ &\leq v^{*T}(x)v^*(x) + g^2_{max}(x) \end{aligned} \tag{28}$$

So, it can be obtained from (26)–(28) that

$$\begin{aligned} \frac{dV^*}{dt} \leq &- \beta^2 x^T x + 2v^{*T}(x)v^*(x) \\ =&2v^{*T}(x)v^*(x) - \beta'^2 x^T x - (\beta^2 - \beta'^2)x^T x \\ \leq &- (\beta^2 - \beta'^2)x^T x \end{aligned}$$

Therefore, by Lyapunov stability theory, the optimal control $u^*(x) = -\frac{1}{2}g^T(x)\nabla V^*$ can make the closed-loop uncertain nonlinear system asymptotically stable. Thus, for a constant $c > 0$, there is a neighborhood $N = \{x : \|x\| < c\}$ such that if the state $x(t)$ enters the neighborhood N, then $x \to 0$ when $t \to \infty$. However, $x(t)$ cannot stay out of the domain N forever; otherwise, for all $t > 0$, there is $\|x(t)\| \geq c$. This implies that

$$\begin{aligned} V^*[x(t)] - V^*[x(0)] &= \int_0^t \dot{V}^*(x(\tau))dt \\ &\leq \int_0^t -(\beta^2 - \beta'^2)x^T x dt \\ &\leq \int_0^t -(\beta^2 - \beta'^2)c^2 dt \\ &= -(\beta^2 - \beta'^2)c^2 t \end{aligned}$$

Hence, when $t \to \infty$, $V^*[x(t)] \leq V^*[x(0)] - (\beta^2 - \beta'^2)c^2 t \to -\infty$. This contradicts the positivity of $V^*[x(t)]$. Therefore, system (16) is globally asymptotically stable. We complete the proof. □

5. Neural Networks Approximation in PI Algorithm

In the first two sections, the robust control of uncertain nonlinear systems was transformed into solving the optimal control of an auxiliary system. However, whether the uncertain system is matched or mismatched, the key issue is how to obtain the solution to corresponding HJB equation. As is well known, it is a nonlinear partial differential equation that is hard to solve. Moreover, solving the HJB equation may lead to the curse of dimensionality [21]. In this section, an online PI algorithm is used to solve the HJB equation iteratively, and neural networks are utilized to approximate the optimal cost in PI algorithm.

5.1. PI Algorithms for Robust Control

For the system with matched uncertainty, the optimal control problem (4) with (5) is considered. For any admissible control, the corresponding cost function can be expressed as

$$\begin{aligned} V[x(t)] &= \int_t^\infty [f^2_{max}(x) + x^T x + u^T u]dt \\ &= \int_t^{t+T} [f^2_{max}(x) + x^T x + u^T u]dt + \int_{t+T}^\infty [f^2_{max}(x) + x^T x + u^T u]dt \end{aligned}$$

where $T > 0$ is a selected constant. Therefore, it follows that

$$V[x(t)] = \int_t^{t+T} [f^2_{max}(x) + x^T x + u^T u]dt + V[x(t+T)] \tag{29}$$

Based on the integral reinforcement relationship (29) and optimal control (9), the PI algorithm of robust control for matched uncertain nonlinear systems is given below.

The convergence of Algorithm 1 is illustrated as follows. The following conclusion gives an equivalent form of the Bellman Equation (30).

Algorithm 1 PI algorithm of robust control for matched uncertain nonlinear systems

(1) Select supremum $f_{max}(x)$ to satisfy $\|h(x)\| \leq f_{max}(x)$;
(2) Initialization: for the nominal nonlinear system (4), select an initial stabilization control $u_0(x)$;
(3) Policy evaluation: for control input $u_i(x)$, calculate cost $V_i(x)$ from the Bellman equation

$$V_i[x(t)] = \int_t^{t+T} [f_{max}^2(x) + x^T x + u_i^T(x)u_i(x)]dt + V_i[x(t+T)] \tag{30}$$

(4) Policy improvement: compute the control law $u_{i+1}(x)$ using

$$u_{i+1}(x) = -\frac{1}{2}g^T(x)\nabla V_i. \tag{31}$$

By repeatedly iterating between (30) and (31), until the control input is convergent.

Proposition 1. *Suppose that $u_i(x)$ is a stabilization controller of nominal system (4). Then the optimal cost $V_i(x)$ solved from (30) is equivalent to solving the following equation*

$$f_{max}^2(x) + x^T x + u_i^T(x)u_i(x) + \nabla V_i[f(x) + g(x)u_i(x)] = 0. \tag{32}$$

Proof. Dividing both sides of (30) by T and finding the limit yields

$$\lim_{T\to 0} \frac{V_i[x(t+T)] - V_i[x(t)]}{T} + \lim_{T\to 0} \frac{\int_t^{t+T} [f_{max}^2(x) + x^T x + u_i^T(x)u_i(x)]dt}{T} = 0$$

From the definition of function limit and L'Hospital's rule, we can get

$$\frac{dV_i[x(t)]}{dt} + \lim_{T\to 0} \frac{d}{dT} \int_t^{t+T} [f_{max}^2(x) + x^T x + u_i^T(x)u_i(x)]dt = 0 \tag{33}$$

It follows that

$$f_{max}^2(x) + x^T x + u_i^T(x)u_i(x) + \nabla V_i[f(x) + g(x)u_i(x)] = 0$$

Thus, we can deduce (32) from (30). On the other hand, along the stable system $\dot{x} = f(x) + g(x)u_i(x)$, finding the time derivative of $V_i(x)$ yields

$$\frac{dV_i[x(t)]}{dt} = \nabla V_i[f(x) + g(x)u_i(x)]$$

Integrating both sides from t to $t + T$, yields

$$V_i[x(t+T)] - V_i[x(t)] = \int_t^{t+T} \nabla V_i[f(x) + g(x)u_i(x)]dt$$

Therefore, we can get the following result from (32)

$$V_i[x(t)] = \int_t^{t+T} [f_{max}^2(x) + x^T x + u_i^T(x)u_i(x)]dt + V_i[x(t+T)]$$

This proves that (32) can deduce (30). □

According to [32–34], if the initial stabilization control policy is given $u_0(x)$, then the follow-up control policy calculated by the iterative relations of (30) and (31) is also a stabilizing control policy, and cost sequence $V_i[x(t)]$ calculated by iteration converges to the optimal cost. By Proposition 1, it is known that (30) and (32) are equivalent, so the iterative relations (30) and (31) in Algorithm 1 converge to the optimal control and optimal cost.

Similarly, we give a PI algorithm of robust control for nonlinear systems with mismatched uncertainties.

The steps of policy evaluation (34) and policy improvement (35) are iteratively calculated until the policy improvement step does not change the current policy. The optimal cost function is calculated as $V^*(x)$, then $u^*(x) = -\frac{1}{2}g^T(x)\nabla V^*(x)$ is the robust control law.

The convergence proof of Algorithm 2 is similar to Algorithm 1, which will not be repeated here.

Algorithm 2 PI algorithm of robust control for nonlinear systems with mismatched uncertainties

(1) Decompose the uncertainty properly so that $\Delta f(x) = c(x)h(x)$ and $\Delta g(x) = g(x)m(x)$, select constant parameter β, β' such that $|\beta'| < |\beta|$, and then calculate the nonnegative function $f_{max}(x)$ and $g_{max}(x)$ according to (19);

(2) For auxiliary system (17), select an initial stabilization control policy $u_0(x)$;

(3) Policy evaluation: Give a control policy $u_i(x)$, the cost $V_i(x)$ is solved from the following Bellman equation

$$V_i[x(t)] = \int_t^{t+T} [f_{max}^2(x) + g_{max}^2(x) + \beta x^T x + \bar{u}_i^T(x)\bar{u}_i(x)]dt + V_i[x(t+T)]; \tag{34}$$

(4) Policy improvement: Calculate the control policy using the following update law

$$\bar{u}_{i+1}(x) = -\frac{1}{2}\bar{g}^T(x)\nabla V_i; \tag{35}$$

(5) Check if the condition $2v^{*T}(x)v^*(x) \leq \beta'^2 x^T x$ is satisfied. Return to step (1) and select the larger constants β and β' when it does not hold.

Remark 3. *In Step (3) of Algorithm 1 or Algorithm 2, solving $V_i[x(t)]$ from (30) or (34) can be transformed into a least squares problem [17]. By reading enough data online along the system trajectory, the cost function $V(x)$ can be calculated by using the least square principle. However, the cost $V_i[x(t)]$ has no specific expressions. In next subsection, along the system trajectory, online reading of sufficient data on the interval $[t, t+T]$, the cost $V_i[x(t)]$ can be approximated by neural network in PI algorithms. Moreover, implementation of the algorithm does not need to know the system dynamics function $f(x)$.*

5.2. Neural Network Approximation of Optimal Cost in PI Algorithm

In the implementation of the PI algorithms, we need to use the data of the nominal system and use the least square method to solve the cost function. However, the cost function of nonlinear optimal control problem has no specific form. Therefore, it is necessary to use neural network structure to approximate the cost function, carry out policy iteration, update weights, and then obtain the approximate optimal cost function. In this subsection, neural network is utilized to approximate the optimal cost in the corresponding HJB equation.

Based on the continuous approximation theory of neural network [35], a single neural network is utilized to approximate the optimal cost in HJB equation. For matched uncertain systems, suppose that the solution $V^*(x)$ of HJB Equation (10) is smooth positive definite, and the optimal cost function on compact set Ω is expressed as

$$V^*(x) = W^T\phi(x) + \varepsilon(x) \tag{36}$$

where $W \in \mathbb{R}^L$ is an unknown ideal weight, and $\phi(.) : \mathbb{R}^n \to \mathbb{R}^L$ is a linear independent basis vector function. It is assumed that $\phi(x)$ is continuous, $\phi(0) = 0$, and $\varepsilon(x)$ is the error vector of neural network reconstruction. Thus, the gradient of the optimal cost function (36) can be expressed as

$$\nabla V^* = \frac{\partial V^*}{\partial x} = \nabla \phi^T(x) W + \nabla \varepsilon(x) \tag{37}$$

where $\nabla \varepsilon(x) = \frac{\partial \varepsilon}{\partial x}$. On the basis of approximation property of neural network [35,36], when the number of neurons in hidden layer $L \to \infty$, the approximation error $\varepsilon(x) \to 0$, $\nabla \varepsilon(x) \to 0$. Substituting (36) and (37) into (9), the optimal control is rewritten as follows

$$u^*(x) = -\frac{1}{2} g^T(x)[\nabla \phi^T(x) + \nabla \varepsilon(x)] \tag{38}$$

Assume that $\hat{W}$ is an estimated value of the ideal weight W. Since the ideal weight W in (36) is unknown, the cost function of the $i-th$ iteration in Algorithm 1 is expressed as

$$\hat{V}_i(x) = \hat{W}_i^T \phi(x) \tag{39}$$

Using the approximation of neural network in cost function, the Bellman Equation (30) in Algorithm 1 is rewritten as follows

$$\hat{W}_i^T \phi(x(t)) = \Psi + \hat{W}_i^T \phi(x(t+T)) \tag{40}$$

where $\Psi = \int_t^{t+T} [f_{max}^2(x) + x^T x + u_i^T(x) u_i(x)] dt$. Since the above formula uses neural network to approximate the cost function, the residual error caused by neural network approximation is

$$\varepsilon_i(x(t), T) = \Psi + \hat{W}_i^T \phi(x(t+T)) - \hat{W}_i^T \phi(x(t)) \tag{41}$$

In order to obtain the neural network weight parameters of approximation function, the following objective functions can be minimized in the meaning of least square

$$E = \int_\Omega \varepsilon_i(x(t), T)^T \varepsilon_i(x(t), T) dx, \tag{42}$$

that is $\int_\Omega \frac{d\varepsilon_i(x(t),T)}{d\hat{W}_i} \varepsilon_i(x(t), T) dx = 0$. Using the definition of inner product, it can be rewritten as

$$\langle \frac{d\varepsilon_i(x(t), T)}{d\hat{W}_i}, \varepsilon_i(x(t), T) \rangle_\Omega = 0 \tag{43}$$

It follows from properties of the internal product that

$$\Phi \hat{W}_i + \langle [\phi(x(t+T)) - \phi(x(t))], \Psi \rangle_\Omega = 0 \tag{44}$$

where $\Phi = \langle [\phi(x(t+T)) - \phi(x(t))], [\phi(x(t+T)) - \phi(x(t))]^T \rangle$. Therefore,

$$\hat{W}_i = -\Phi^{-1} \langle [\phi(x(t+T)) - \phi(x(t))], \Psi \rangle_\Omega \tag{45}$$

So far, the neural network weight parameters of approximation function $V_i(x)$ can be calculated. Thus, the update control policy can be obtained from (35)

$$\hat{u}_{i+1}(x) = -\frac{1}{2} g^T(x) \nabla \phi^T(x) \hat{W}_i. \tag{46}$$

According to [32,33,35,36], using the policy iteration of RL algorithm, the cost sequence $V_i(x)$ converges to the optimal cost $V^*(x)$, and the control sequence $u_i(x)$ converges to the optimal control function $u^*(x)$.

For mismatched uncertain systems, similar neural network approximation can be used.

6. Simulation Examples

Some simulation examples are presented to verify the feasibility of the robust control design method for uncertain nonlinear systems in this section.

Example 1. *Consider the following uncertain nonlinear systems*

$$\dot{x}(t) = \begin{bmatrix} 0 & 6 \\ -1 & 1 \end{bmatrix} \begin{bmatrix} x_1 \\ x_2 \end{bmatrix} + \begin{bmatrix} 0 \\ p_1 x_1 cos(x_2^2) \end{bmatrix} + \begin{bmatrix} 0 \\ 1 + p_2 x_2^2 \end{bmatrix} u \tag{47}$$

where $x = [x_1, x_2]^T$ *is the system state,* $\Delta f(x) = \begin{bmatrix} 0 \\ p_1 x_1 cos(x_2^2) \end{bmatrix}$ *is the uncertain disturbance function of the system,* $\Delta g(x) = \begin{bmatrix} 0 \\ p_2 x_2^2 \end{bmatrix}$ *is input uncertainty function,* $p_1 \in [-2, 2]$, $p_2 \in [0, 10]$.

Obviously,

$$\Delta f(x) = g(x)h(x), \Delta g(x) = g(x)m(x) \tag{48}$$

where $g(x) = \begin{bmatrix} 0 \\ 1 \end{bmatrix}$, $h(x) = p_1 x_1 cos(x_2^2)$, $m(x) = p_2 x_2^2$. *Moreover,* $|h(x)| \leq |2x_1| = f_{max}(x)$. *Thus, the original robust control problem is converted into calculating optimal control law. For nominal system*

$$\dot{x}(t) = \begin{bmatrix} 0 & 6 \\ -1 & 1 \end{bmatrix} \begin{bmatrix} x_1 \\ x_2 \end{bmatrix} + \begin{bmatrix} 0 \\ 1 \end{bmatrix} u, \tag{49}$$

find the control function u, such that the performance index

$$\begin{aligned} J(x, u) &= \int_0^{\infty} [f_{max}^2(x) + x^T x + u^T u] dt \\ &= \int_0^{\infty} [5x_1^2 + x_2^2 + u^2] dt \end{aligned} \tag{50}$$

is minimized.

In order to solve the robust control problem by using Algorithm 1, it is assumed that the optimal cost function $V^*(x)$ *has a neural network structure:* $V^*(x) = W^T \phi(x)$, *where* $W = [W_1, W_2, W_3]^T$, $\phi(x) = [x_1^2, x_1 x_2, x_2^2]^T$. *The initial weight is taken as* $W_0 = [-1, 5, 1.5]^T$, *and the initial state of system* $x_0 = [2, -0.5]^T$. *The neural network weights are calculated iteratively by MATLAB. In each iteration, 10 sets of data samples are collected along the nominal system trajectory to perform the batch least squares problem. After five iterations, the weight converges to* $[1.9645, 2.8990, 5.4038]$. *The robust control law of uncertain system (47) is* $u^* = -1.4495x_1 - 5.4038x_2$. *The convergence process of neural network weight is shown in Figure 1, while the changing process of control signal is shown in Figure 2. The uncertain parameter* p_1 *and* p_2 *in uncertain system (47) take different values, the state trajectories of the closed-loop system are obtained by the robust control law. Figure 3 shows the trajectory of the closed-loop system when* $p_1 = -2$, $p_2 = 1$. *Figure 4 shows the trajectory of the closed-loop system when* $p_1 = -1$, $p_2 = 4$. *Figure 5 shows the trajectory of the closed-loop system when* $p_1 = 0$, $p_2 = 7$. *Figure 6 shows the trajectory of the closed-loop system is* $p_1 = 2$, $p_2 = 10$. *From these figures, we can see that the closed-loop system is stable, which shows the effectiveness of the robust control law.*

In this example, because of the linear property of the nominal system, MATLAB software can be used to solve LQR problem directly. With this method, the optimal control is calculated as $u^* = -1.4496x_1 - 5.4038x_2$. *It is almost the same as the result of neural network approximation, which shows the validity of Algorithm 1.*

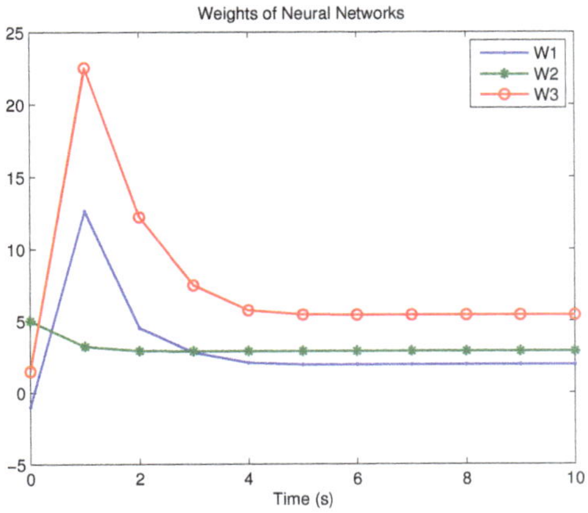

Figure 1. Neural network weight.

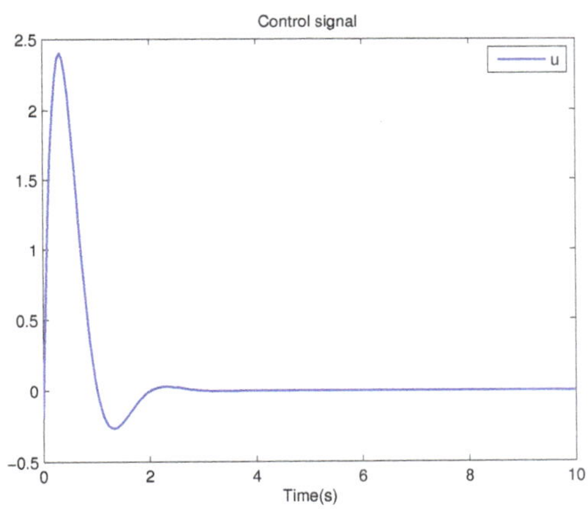

Figure 2. Robust control signal.

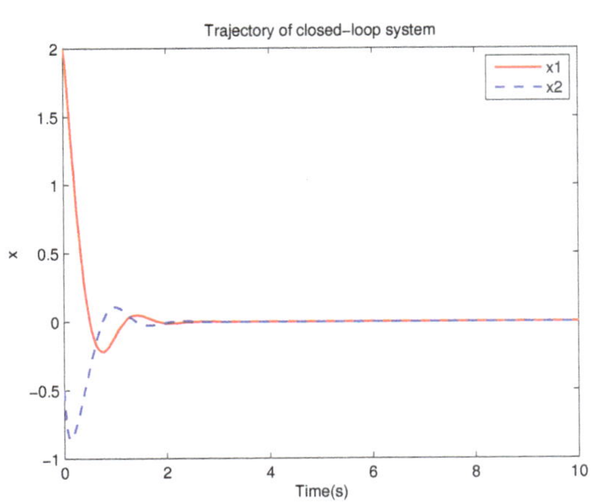

Figure 3. Closed loop system trajectory, $p_1 = -2, p_2 = 1$.

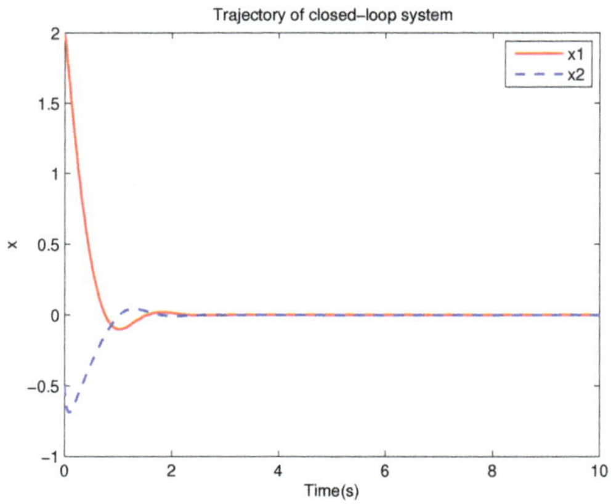

Figure 4. Closed loop system trajectory, $p_1 = -1, p_2 = 4$.

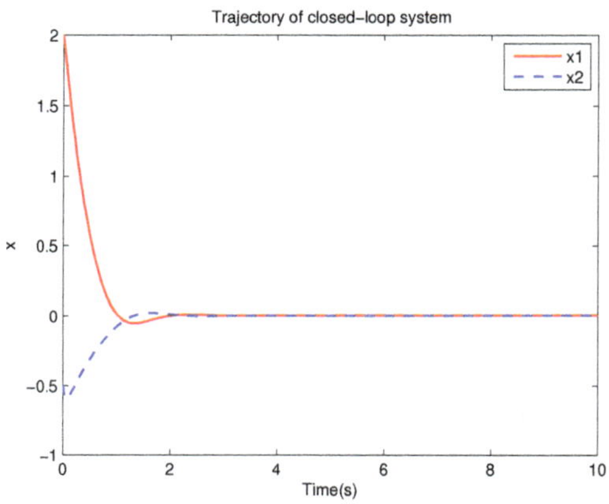

Figure 5. Closed loop system trajectory, $p_1 = 0, p_2 = 7$.

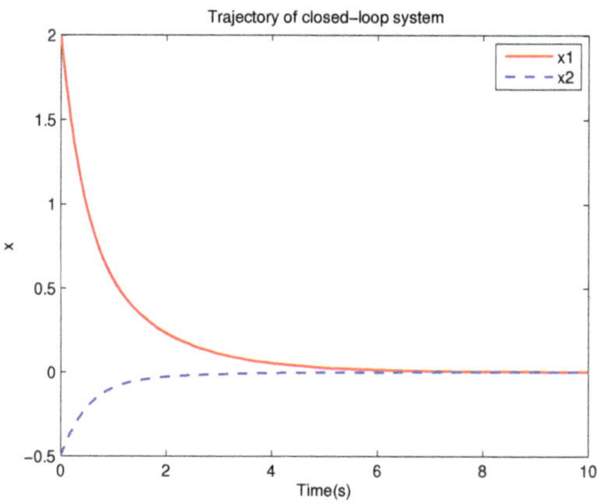

Figure 6. Closed loop system trajectory, $p_1 = 2, p_2 = 10$.

Example 2. *Consider the following uncertain nonlinear systems*

$$\dot{x}(t) = \begin{bmatrix} 0 & 8 \\ -5 & 1 \end{bmatrix}\begin{bmatrix} x_1 \\ x_2 \end{bmatrix} + \begin{bmatrix} 0 \\ 0.5 + p_2x_2^2 \end{bmatrix}u + \begin{bmatrix} p_1x_1cos(x_2^2) + p_3x_2sin(x_1x_2) \\ 0 \end{bmatrix} \quad (51)$$

where $x = [x_1, x_2]^T$ *is the system state,* $p_1 \in [-2,2]$, $p_2 \in [0,5]$, $p_3 \in [-1,1]$. *Let* $\Delta f(x) = \begin{bmatrix} p_1x_1cos(x_2^2) + p_3x_2sin(x_1x_2) \\ 0 \end{bmatrix}$, $\Delta g(x) = \begin{bmatrix} 0 \\ p_2x_2^2 \end{bmatrix}$. *It is easy to know that the system (51) is a mismatched system. The uncertain disturbance of the system is decomposed as*

$$\Delta f(x) = c(x)h(x), \Delta g(x) = g(x)m(x) \quad (52)$$

where, $g(x) = \begin{bmatrix} 0 \\ 0.5 \end{bmatrix}$, $c(x) = \begin{bmatrix} 1 \\ 0 \end{bmatrix}$, $h(x) = p_1x_1cos(x_2^2) + p_3x_2sin(x_1x_2)$, $m(x) = p_2x_2^2$. *Moreover,* $f_{max}(x)$ *and* $g_{max}(x)$ *are calculated as follows.*

$$\|g(x)^+c(x)h(x)\| = \|\begin{bmatrix} 0 & 2 \end{bmatrix}\begin{bmatrix} 1 \\ 0 \end{bmatrix}h(x)\| = 0 = f_{max}(x),$$

and

$$\|h(x)\| = \|p_1x_1cos(x_2^2) + p_3x_2sin(x_1x_2)\| \leq |2x_1 + x_2| = g_{max}(x)$$

Select the parameter $\beta = 1$. *Then the original robust control problem is converted into solving an optimal control problem. For the auxiliary system*

$$\dot{x}(t) = \begin{bmatrix} 0 & 8 \\ -5 & 1 \end{bmatrix}\begin{bmatrix} x_1 \\ x_2 \end{bmatrix} + \begin{bmatrix} 0 & 1 \\ 0.5 & 0 \end{bmatrix}\bar{u}, \quad (53)$$

find the control policy, $\bar{u}$, *such that the following performance index is minimized*

$$\begin{aligned} J(x,u) &= \int_0^\infty [f_{max}^2(x) + g_{max}^2(x) + \beta^2x^Tx + \bar{u}^T\bar{u}]dt \\ &= \int_0^\infty [5x_1^2 + 2x_2^2 + 4x_1x_2 + \bar{u}^T\bar{u}]dt. \end{aligned} \quad (54)$$

In order to obtain the obust control law by using Algorithm 2, it is assumed that the optimal cost function $V^*(x)$ *has a neural network structure:* $V^*(x) = W^T\phi(x)$, *where* $W = [W_1, W_2, W_3]^T$, $\phi(x) = [x_1^2, x_1x_2, x_2^2]^T$. *The initial weight is taken as* $W_0 = [1, -3, 0.5]^T$, *and the initial state of system is chosen as* $x_0 = [-2, 0.5]^T$. *The neural network weights are calculated iteratively by MATLAB. In each iteration, 10 sets of data samples are collected along the nominal system trajectory to perform the batch least squares problem. After six iterations, the weight converges to* $W = [2.8983, -0.6859, 5.2576]^T$. *The optimal control of the auxiliary system is calculated as* $\bar{u}^* = \begin{bmatrix} 0.1715x_1 - 2.6288x_2 \\ -2.8983x_1 + 0.3429x_2 \end{bmatrix}$. *The robust control law of the original uncertain system is* $u^* = 0.1715x_1 - 2.6288x_2$. *The convergence process of neural network weight is shown in Figure 7, while the changing process of control signal is shown in Figure 8. The uncertain parameters* p_1, p_2 *and* p_3 *in uncertain system (51) take different values, the state trajectories of the closed-loop system are obtained by the robust control law. Figure 9 shows the trajectory of the closed-loop system when* $p_1 = -1$, $p_2 = 1$ *and* $p_3 = 1$. *Figure 10 shows the trajectory of the closed-loop system when* $p_1 = -1$, $p_2 = 2$ *and* $p_3 = 0$. *Figure 11 shows the trajectory of the closed-loop system when* $p_1 = 0.3$, $p_2 = 3$ *and* $p_3 = -1$. *Figure 12 shows the trajectory of the closed-loop system when* $p_1 = -2$, $p_2 = 5$ *and* $p_3 = 1$. *From these figures, we can see that the closed-loop system is stable, which shows the effectiveness of the robust control law.*

The nominal system is also a linear system, so MATLAB software can be used to solve LQR problem directly. With this method, the optimal control is calculated as $\bar{u}^* = \begin{bmatrix} 0.1713x_1 - 2.6286x_2 \\ -2.8983x_1 + 0.3430x_2 \end{bmatrix}$. *It has little difference with the approximate result of neural network, which shows the validity of Algorithm 2.*

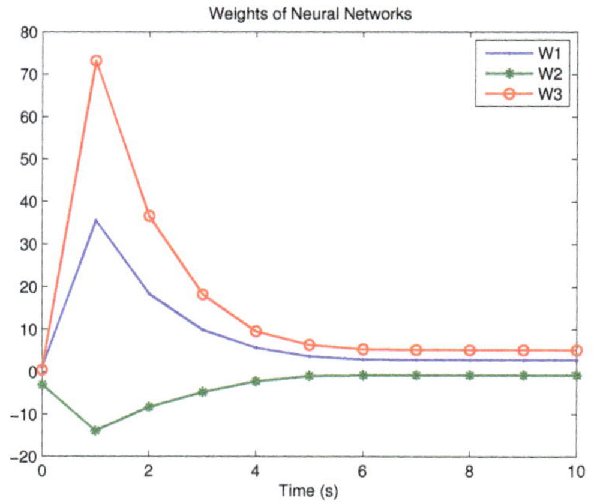

Figure 7. Neural network weight.

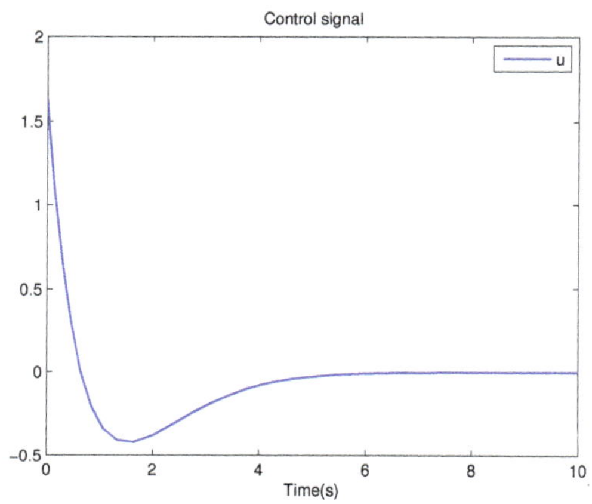

Figure 8. Robust control signal.

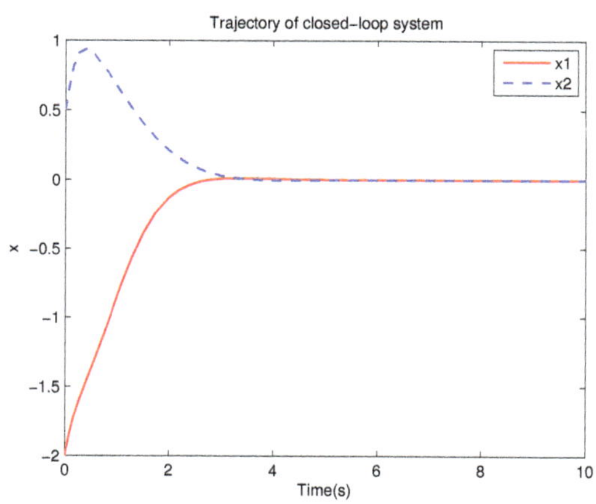

Figure 9. Closed loop system trajectory, $p_1 = -1, p_2 = 2, p_3 = 1$.

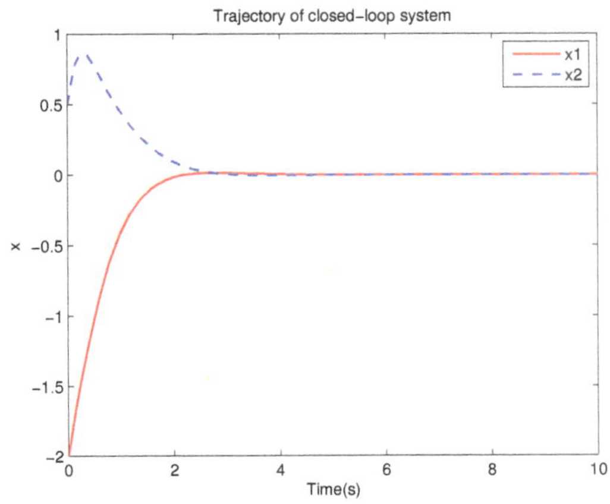

Figure 10. Closed loop system trajectory, $p_1 = -1, p_2 = 2, p_3 = 0$.

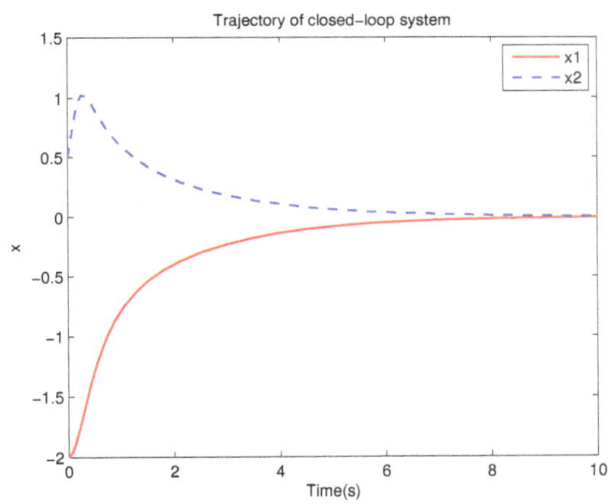

Figure 11. Closed loop system trajectory, $p_1 = 0.3, p_2 = 3, p_3 = -1$.

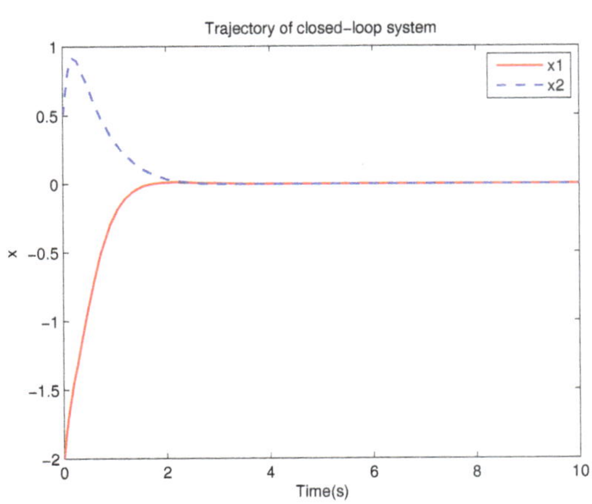

Figure 12. Closed loop system trajectory, $p_1 = -2, p_2 = 5, p_3 = 1$.

The nominal systems corresponding to the above two examples are linear systems. The following is an example with nonlinear nominal system.

Example 3. *Consider the following uncertain nonlinear systems*

$$\dot{x}(t) = \begin{bmatrix} 1 & 1 \\ -1 & -2 \end{bmatrix} \begin{bmatrix} x_1 \\ x_2 \end{bmatrix} + \begin{bmatrix} 0 \\ 2x_1^2 cos^2(x_2) \end{bmatrix} + \begin{bmatrix} 0 \\ p_1 x_2 cos^2(x_1) \end{bmatrix} + \begin{bmatrix} 0 \\ 1 + p_2 x_2^2 \end{bmatrix} u \tag{55}$$

where $x = [x_1, x_2]^T$ *is the system state,* $\Delta f(x) = \begin{bmatrix} 0 \\ p_1 x_2 cos^2(x_1) \end{bmatrix}$ *is the uncertain disturbance function of the system,* $\Delta g(x) = \begin{bmatrix} 0 \\ p_2 x_2^2 \end{bmatrix}$ *is input uncertainty function,* $p_1 \in [-2, 2]$, $p_2 \in [0, 2]$.

Obviously,

$$\Delta f(x) = g(x)h(x), \Delta g(x) = g(x)m(x) \tag{56}$$

where $g(x) = \begin{bmatrix} 0 \\ 1 \end{bmatrix}$, $h(x) = p_1 x_2 cos^2(x_1)$, $m(x) = p_2 x_2^2$. *Moreover,* $|h(x)| \leq |2x_2| = f_{max}(x)$. *Thus, the original robust control problem is converted into calculating optimal control law. For nominal system*

$$\dot{x}(t) = \begin{bmatrix} 1 & 1 \\ -1 & -2 \end{bmatrix} \begin{bmatrix} x_1 \\ x_2 \end{bmatrix} + \begin{bmatrix} 0 \\ 2x_1^2 cos^2(x_2) \end{bmatrix} + \begin{bmatrix} 0 \\ 1 \end{bmatrix} u, \tag{57}$$

find the control function u, such that the performance index

$$\begin{aligned} J(x, u) &= \int_0^{\infty} [f_{max}^2(x) + x^T x + u^T u] dt \\ &= \int_0^{\infty} [x_1^2 + 5x_2^2 + u^2] dt \end{aligned} \tag{58}$$

is minimized.

In order to solve the robust control problem by using Algorithm 1, it is assumed that the optimal cost function $V^*(x)$ *has a neural network structure:* $V^*(x) = W^T \phi(x)$, *where* $W = [W_1, W_2, W_3]^T$, $\phi(x) = [x_1^2, x_1 x_2, x_2^2]^T$. *The initial weight is taken as* $W_0 = [-2, 5, 0.5]^T$, *and the initial state of system* $x_0 = [2, -0.5]^T$. *The neural network weights are calculated iteratively by MATLAB. In each iteration, 10 sets of data samples are collected along the nominal system trajectory to perform the batch least squares problem. After five iterations, the weight converges to* $[25.5830, 12.5830, 2.6458]$. *The robust control law of uncertain system (55) is* $u^* = -6.2915x_1 - 2.6458x_2$. *The convergence process of neural network weight is shown in Figure 13, while the changing process of control signal is shown in Figure 14. The uncertain parameter* p_1 *and* p_2 *in uncertain system (55) take different values, the state trajectories of the closed-loop system are obtained by the robust control law. Figure 15 shows the trajectory of the closed-loop system when* $p_1 = 1$, $p_2 = 0.8$. *Figure 16 shows the trajectory of the closed-loop system when* $p_1 = -0.5$, $p_2 = 1$. *Figure 17 shows the trajectory of the closed-loop system when* $p_1 = 1$, $p_2 = 2$. *Figure 18 shows the trajectory of the closed-loop system is* $p_1 = 2$, $p_2 = 1$. *From these figures, we can see that the closed-loop system is stable, which shows the effectiveness of the robust control law.*

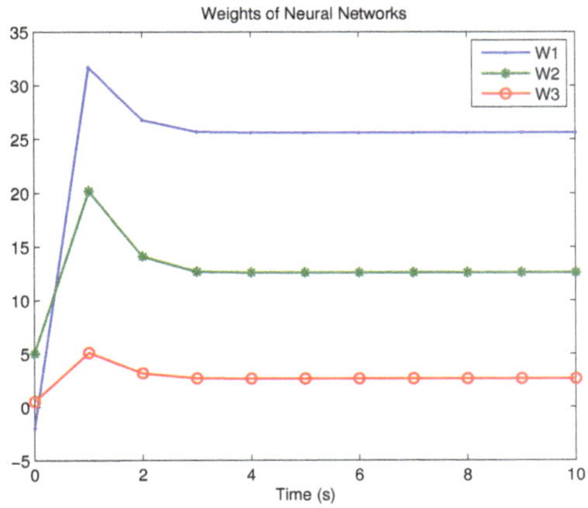

Figure 13. Neural network weight.

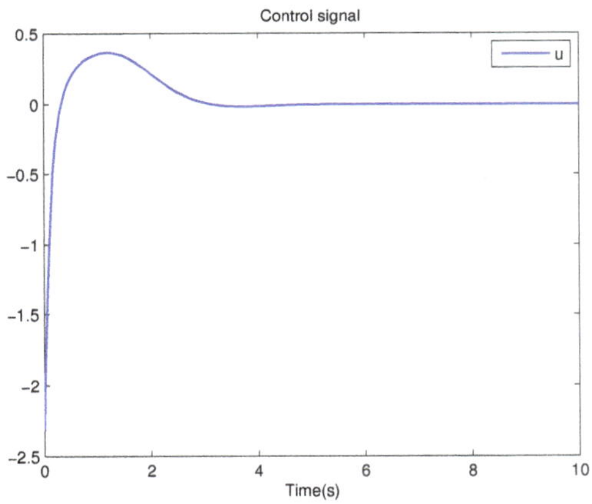

Figure 14. Robust control signal.

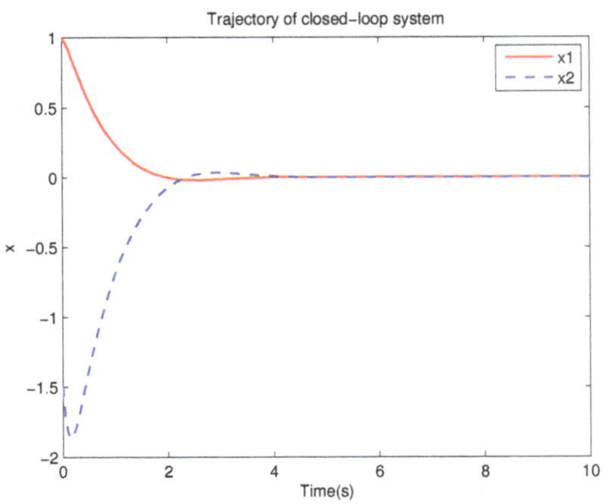

Figure 15. Closed loop system trajectory, $p_1 = 1, p_2 = 0.8$.

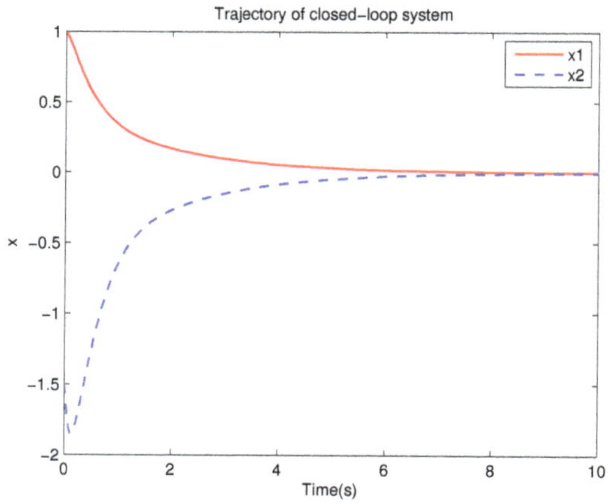

Figure 16. Closed loop system trajectory, $p_1 = -0.5, p_2 = 1$.

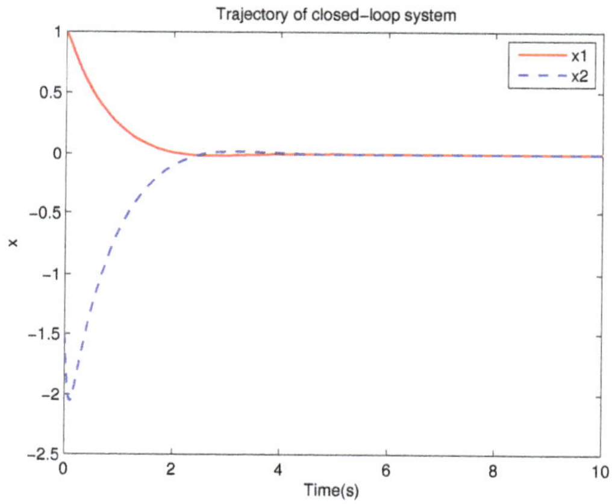

Figure 17. Closed loop system trajectory, $p_1 = 1, p_2 = 2$.

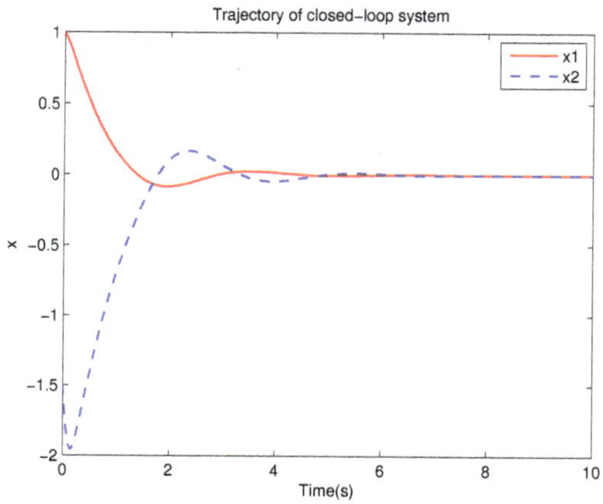

Figure 18. Closed loop system trajectory, $p_1 = 2, p_2 = 1$.

7. Conclusions

In this paper, the PI algorithms in RL are proposed to solve robust control problem for a class of nonlinear continuous time uncertain system. The robust control law is obtained without knowing the internal dynamics of the nominal system. The considered robust control problem is converted into solving an optimal control problem containing a nominal or auxiliary system with a predefined performance index. The online PI algorithms are established to calculate the robust controller of matched and mismatched system. The numerical examples are given to show the availability of the theoretical results. The proposed method may be extended to solve robust tracking problems for some nonlinear systems with uncertainty entering output, which may be the subject of our future research.

Author Contributions: D.X.: investigation, methodology, software; Q.W.: formal analysis; Y.L.: investigation; All the authors contributed equally to the development of the research. All authors have read and agreed to the published version of the manuscript.

Funding: This research was funded by the National Natural Science Foundation of China under Grant No. 61463002.

Institutional Review Board Statement: Not applicable.

Informed Consent Statement: Not applicable.

Data Availability Statement: Data is contained within the article.

Acknowledgments: The authors thank to the Journal editors and the reviewers for their helpful suggestions and comments.

Conflicts of Interest: The authors declare no conflict of interest.

References

1. Zhou, K.; Doyle, J.C.; Glover, K. *Robust and Optimal Control*; Prentice Hall: Englewood Cliffs, NJ, USA, 1996.
2. Petersen, I.R.; Hollot, C.V. A riccati equation approach to the stabilization of uncertain linear systems. *Automatica* **1986**, *22*, 397–411. [CrossRef]
3. Schmitendorf, W.E. Designing stabilizing controllers for uncertain systems using the riccati equation approach. *IEEE Trans. Autom. Control* **1988**, *33*, 376–379. [CrossRef]
4. Agulhari, C.M.; Oliveira, R.C.; Peres, P.L. Relaxations for Reduced-Order Robust H_∞ Control of Continuous-Time Uncertain Linear Systems. *IEEE Trans. Autom. Control* **2012**, *57*, 1532–1537. [CrossRef]
5. Jabbari, F.; Schmitendorf, W. A noniterative method for the design of linear robust controllers. *IEEE Trans. Autom. Control* **1990**, *35*, 954–957. [CrossRef]
6. Tsay, S.C. Robust control for linear uncertain systems via linear quadratic state feedback. *Syst. Control Lett.* **1990**, *15*, 199–205. [CrossRef]
7. Marino, R.; Tomei, P. Robust stabilization of feedback linearizable time-varying uncertain nonlinear systems. *Automatica* **1993**, *29*, 181–189. [CrossRef]
8. Shen, T.; Tamura, K. Robust H_∞ control of uncertain nonlinear system via state feedback. *IEEE Trans. Autom. Control* **2002**, *40*, 766–768. [CrossRef]
9. Teixeira, M.C.M.; Zak, S.H. Stabilizing controller design for uncertain nonlinear systems using fuzzy models. *IEEE Trans. Fuzzy Syst.* **1999**, *7*, 133–142. [CrossRef]
10. Roy, S.; Kar, I.N.; Lee, J.; Jin, M. Adaptive-Robust Time-Delay Control for a Class of Uncertain Euler–Lagrange Systems. *IEEE Trans. Ind. Electron.* **2017**, *64*, 7109–7119. [CrossRef]
11. Hosseinzadeh, M.; Yazdanpanah, M.J. Performance enhanced model reference adaptive control through switching non-quadratic Lyapunov functions. *Syst. Control Lett.* **2015**, *76*, 47–55. [CrossRef]
12. Ma, J.; Xu, S.; Zhuang, G.; Wei, Y.; Zhang, Z. Adaptive neural network tracking control for uncertain nonlinear systems with input delay and saturation. *Int. J. Robust Nonlinear Control* **2020**, *30*, 2593–2610. [CrossRef]
13. Zhou, L.; She, J.; Zhang, X.M.; Zhang, Z. Additive-state-decomposition based repetitive-control framework for a class of nonlinear systems with multiple mismatched disturbances. *IEEE Trans. Ind. Electron.* **2020**, to be published. [CrossRef]
14. Liu, L.; Liu, Y.J.; Li, D.; Tong, S.; Wang, Z. Barrier Lyapunov function based adaptive fuzzy FTC for switched systems and its applications to resistance inductance capacitance circuit system. *IEEE Trans. Cybern.* **2020**, *50*, 3491–3502. [CrossRef]
15. Werbos, P.J. Approximate dynamic programming for real-time control and neural modeling. In *Handbook of Intelligent Control: Neural, Fuzzy, and Adaptive Approaches*; Van Nostrand Reinhold Company: New York, NY, USA, 1992; pp. 493–526.
16. Bhasin, S.; Sharma, N.; Patre, P.; Dixon, W. Asymptotic tracking by a reinforcement learning-based adaptive critic controller. *J. Control Theory Appl.* **2011**, *9*, 400–409. [CrossRef]

17. Liu, Y.J.; Tang, L.; Tong, S.; Chen, C.P.; Li, D.J. Reinforcement learning design-based adaptive tracking control with less learning parameters for nonlinear discrete-time mimo systems. *IEEE Trans. Neural Netw. Learn. Syst.* **2014**, *26*, 165–176. [CrossRef] [PubMed]
18. Vrabie, D.; Pastravanu, O.; Abu-Khalaf, M.; Lewis, F.L. Adaptive optimal control for continuous-time linear systems based on policy iteration. *Automatica* **2009**, *45*, 477–484. [CrossRef]
19. Modares, H.; Lewis, F.L. Linear quadratic tracking control of partially-unknown continuous-time systems using reinforcement learning. *IEEE Trans. Autom. Control* **2014**, *59*, 3051–3056. [CrossRef]
20. Kiumarsi, B.; Vamvoudakis, K.G.; Modares, H.; Lewis, F.L. Optimal and autonomous control using reinforcement learning: A survey. *IEEE Trans. Neural Netw. Learn. Syst.* **2018**, *29*, 2042–2062. [CrossRef] [PubMed]
21. Lin, F. An optimal control approach to robust control design. *Int. J. Control* **2000**, *73*, 177–186. [CrossRef]
22. Wang, D.; He, H.; Liu, D. Adaptive critic nonlinear robust control: A survey. *IEEE Trans. Cybern.* **2017**, *47*, 3429–3451. [CrossRef]
23. Bhasin, S.; Kamalapurkar, R.; Johnson, M.; Vamvoudakis, K.G.; Lewis, F.L.; Dixon, W.E. A novel actor-critic-identifier architecture for approximate optimal control of uncertain nonlinear systems. *Automatica* **2013**, *49*, 82–92. [CrossRef]
24. Jiang, H.; Zhang, H. Robust control scheme for a class of uncertain nonlinear systems with completely unknown dynamics using data-driven reinforcement learning method. *Neurocomputing* **2018**, *273*, 68–77. [CrossRef]
25. Zhang, Q.; Zhao, D.; Wang, D. Event-based robust control for uncertain nonlinear systems using adaptive dynamic programming. *IEEE Trans. Neural Netw. Learn. Syst.* **2018**, *29*, 37–50. [CrossRef] [PubMed]
26. Yang, X.; He, H.; Zhong, X. Adaptive dynamic programming for robust regulation and its application to power systems. *IEEE Trans. Ind. Electron.* **2018**, *65*, 5722–5732. [CrossRef]
27. Jia, S.; Jiang, Y.; Li, T.; Du, Y. Learning-Based Optimal Desired Compensation Adaptive Robust Control for a Flexure-Based Micro-Motion Manipulator. *Appl. Sci.* **2017**, *7*, 406. [CrossRef]
28. Lewis, F.L.; Vrabie, D.; Syrmos, V.L. *Syrmos, Optimal Control;* John Wiley & Sons: New York, NY, USA, 2012.
29. Lin, F. *Robust Control Design: An Optimal Control Approach;* John Wiley & Sons: New York, NY, USA, 2007.
30. Sidorov, N.; Sidorov, D.; Sinitsyn, A.V. *Toward General Theory of Differential-Operator and Kinetic Models;* World Scientific: Singapore, 2020.
31. Ben-Israel, A.; Greville, T.N.E. *Generalized Inverses: Theory and Applications;* Springer Science & Business Media: New York, NY, USA, 2003.
32. Abu-Khalaf, M.; Lewis, F.L. Nearly optimal control laws for nonlinear systems with saturating actuators using a neural network HJB approach. *Automatica* **2005**, *41*, 779–791. [CrossRef]
33. Vrabie, D.; Lewis, F. Neural network approach to continuous-time direct adaptive optimal control for partially unknown nonlinear systems. *Neural Netw.* **2009**, *22*, 237–246. [CrossRef] [PubMed]
34. Xu, D.; Wang, Q.; Li, Y. Optimal Guaranteed Cost Tracking of Uncertain Nonlinear Systems Using Adaptive Dynamic Programming with Concurrent Learning. *Int. J. Control Autom. Syst.* **2020**, *18*, 1116–1127. [CrossRef]
35. White, H. *Artificial Neural Networks: Approximation and Learning Theory;* Blackwell Publishers, Inc.: Cambridge, UK, 1992.
36. Zhou, Q.; Shi, P.; Tian, Y.; Wang, M. Approximation-Based Adaptive Tracking Control for MIMO Nonlinear Systems with Input Saturation. *IEEE Trans. Cybern.* **2015**, *45*, 2119–2128. [CrossRef] [PubMed]

applied sciences

MDPI

Article

Digital Twin for Designing and Reconfiguring Human–Robot Collaborative Assembly Lines

Niki Kousi, Christos Gkournelos, Sotiris Aivaliotis, Konstantinos Lotsaris, Angelos Christos Bavelos, Panagiotis Baris, George Michalos and Sotiris Makris *

Laboratory for Manufacturing Systems and Automation, University of Patras, 26504 Patras, Greece; kousi@lms.mech.upatras.gr (N.K.); gkournelos@lms.mech.upatras.gr (C.G.); saival@lms.mech.upatras.gr (S.A.); lotsaris@lms.mech.upatras.gr (K.L.); bavelos@lms.mech.upatras.gr (A.C.B.); pmparis@lms.mech.upatras.gr (P.B.); michalos@lms.mech.upatras.gr (G.M.)

* Correspondence: makris@lms.mech.upatras.gr; Tel.: +30-2610-910160

Abstract: This paper discusses a digital twin-based approach for designing and redesigning flexible assembly systems. The digital twin allows modeling the parameters of the production system at different levels including assembly process, production station, and line level. The approach allows dynamically updating the digital twin in runtime, synthesizing data from multiple 2D–3D sensors in order to have up-to-date information about the actual production process. The model integrates both geometrical information and semantics. The model is used in combination with an artificial intelligence logic in order to derive alternative configurations of the production system. The overall approach is discussed with the help of a case study coming from the automotive industry. The case study introduces a production system integrating humans and autonomous mobile dual arm workers.

Keywords: digital twin; human robot collaboration; reconfiguration

Citation: Kousi, N.; Gkournelos, C.; Aivaliotis, S.; Lotsaris, K.; Bavelos, A.C.; Baris, P.; Michalos, G.; Makris, S. Digital Twin for Designing and Reconfiguring Human–Robot Collaborative Assembly Lines. *Appl. Sci.* **2021**, *11*, 4620. https://doi.org/10.3390/app11104620

Academic Editor: Emanuele Carpanzano

Received: 23 April 2021
Accepted: 15 May 2021
Published: 18 May 2021

1. Introduction

Traditional production systems are facing difficulties in accommodating the market demand for customized products [1]. This is attributed to the rigidity of their structure and control logic [2] as well as the isolation between design and production phases [3]. This has led to a manufacturing trend toward enabling hybrid and more flexible production systems [4].

New types of factories that facilitate the cooperation among multipurpose resources achieving the system's reconfigurability have been emerging in the last decades [2]. The current work aims to enable the deployment of such reconfigurable production systems employing mobile dual arm workers that can navigate across the shopfloor to undertake different operations while acting as assistants to human operators. To facilitate the autonomous and dynamic behavior in those systems, it is critical to implement methods and tools to optimally design and plan the activities among the available humans and robots and re-allocate the work when changes are needed or unexpected events occur [5].

Today's industrial practice lacks a systematic approach to automate the process of designing the end-to-end operation of a manufacturing system [6]. In most of the cases, a manual design is generated by the engineers based on their experience using isolated engineering tools. In case of disturbances during the execution, such as changes in the production mix or the resources' availability, the system setup and the task sequence are manually re-designed, leading to the extended use of manual resources as well as stoppage of the production [7].

Digital manufacturing aspires to bridge this gap between the design and operation phases of a product [8]. The integration of smart Digital Twins (DT) combined with Artificial Intelligence (AI) methods is a key step for the evolution of decision-making systems [9]. Over the last decades, DT based decision-making frameworks have been suggested advancing multiple aspects in Smart Manufacturing: (a) business data integration [10], surveillance and predictive maintenance [11], offline robot work cell design [12] and programming [13], and online robot behavior adaptation [14].

Over the last few years, several definitions for the term "Digital Twin" have been presented [15]. Different simulation-based techniques for the designing and the performance assessment of flexible manufacturing systems have been presented [16,17]. The connection of a DT with the sustainability validation of an intelligent manufacturing system has been highlighted, and a framework to achieve this has been proposed [18]. An ISO standard for designing and implementing DTs is available to help SMEs develop a DT for their use cases [19]. A detailed documentation on the deployment of a DT for discrete manufacturing systems has been provided [20].

Digital Twinning techniques involve the deployment of multiple sensors for virtual reconstructing the real-time status of the factory [3,21]. To capture and process the data coming from multiple sensing devices, various sensor fusion techniques have been developed [22]. Sensor fusion methods in the field of wearable robotics (prostheses and exoskeletons) have been investigated [23]. Additionally, the navigation ability of mobile and aerial robots has been developed based on data fusion techniques [24,25].

The use of DT for the design and control of Human–Robot Collaborative (HRC) operations has also been investigated [26]. However, the existing approaches are limited only in design and planning for HRC systems that incorporate stationary robots. Therefore, they do not consider the use of mobile robots and its effect in layout reconfigurability. Furthermore, there is a lack of end-to-end integrated systems to monitor the workplace and resources status and dynamically re-design the system in case of unexpected events.

This work introduces a DT-based system to close the loop between the design and operation phase enabling the dynamic behavior of HRC reconfigurable systems. The DT is integrated with an AI decision-making logic to design the layout and the task plans of the system. During runtime, the DT smart models are dynamically updated through unified interfaces with the physical assets synthesizing the real-time planning scene that is used when re-design is needed. The goal of the proposed approach is to overcome the existing limitations in terms of execution orchestration and online reconfiguration of hybrid production systems by implementing;

- Simplified Robot Operating System (ROS) [27] based control integration and sensor data sharing through services. This results in an easier distribution of acquired data and tasks to all relevant resources such as mobile robots and human operators via Human–Machine Interfaces, etc.,
- Virtual representation of the shopfloor by combining multiple sensor data with resource-related information and CAD models. The information is continuously updated through a network of services by all resources and sensors creating a synthesis of all perception data, compiling the Digital Twin.
- Integration of two level of decision-making modules: (a) layout and task planning and (b) mobile robots' path and motion planning. These modules consume the virtual representation of the shopfloor toward generating optimal layout and task plans to online reconfigure the system in cases of unexpected events.

The overall concept for the suggested DT-based reconfiguration system is illustrated in Figure 1.

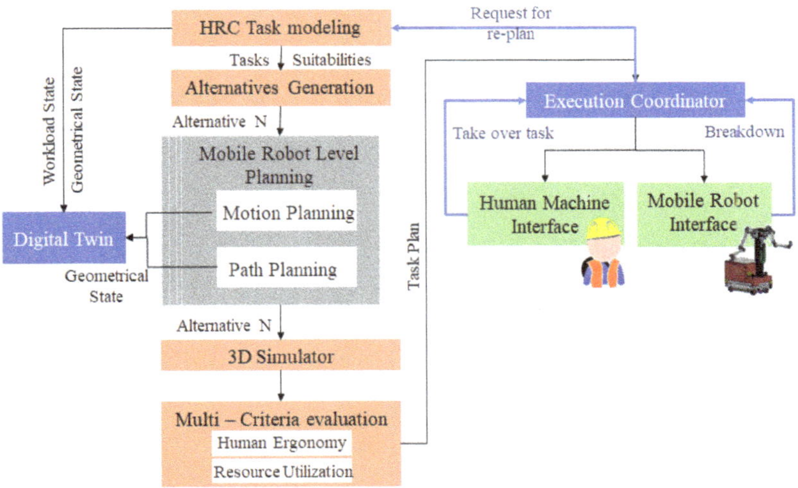

Figure 1. Proposed DT-based reconfiguration concept.

The paper structure comprises an overview of the suggested DT-based re-design process (Section 2) and its implementation for HRC reconfigurable workplaces (Section 3). The performance of the system is analyzed and validated in an automotive case study (Section 4). Finally, Section 5 is dedicated to drawing the conclusions and providing an outlook on future research areas.

2. Approach

The proposed method supports the continuous design in flexible manufacturing systems exploiting digital twin technology. Based on Stark et al.'s DT definition [28], a DT infrastructure, digitalizing multiple assets, is proposed to close the loop between the design tools and the physical world, as shown in Figure 2. Three steps are followed to set up the DT system.

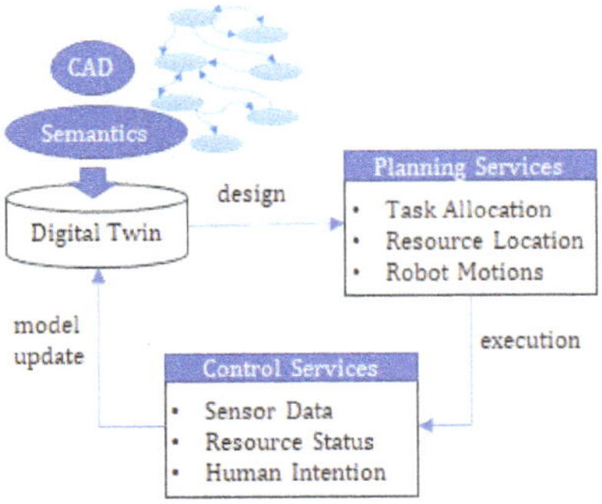

Figure 2. The continuous re-design process.

The first step includes the configuration of the DT during the design phase. Currently, production designers use various simulation tools to design the production system. However, the uncertainty on the shop floor during execution such as human operators' unpredicted behavior or unexpected variations in the demand profile or in the resources' availability is not captured sufficiently through these tools. The suggested DT includes a unified representation of the Product–Process–Resource system aiming to link the production design and execution phases. The configuration process involves the creation of smart digital models to represent multiple assets of the shopfloor, including the parts, processes, resources, and sensors. Having as a starting point the CAD models of the parts and of the production line, these models comprise (a) the semantic representation of each entity including its characteristic and capabilities as well as of the tasks that need to be executed, (b) the geometrical information that defines the spatial relation between the entities, and (c) uniform data structures able to synthesize multiple sensor data types. The target is to use these models for data sharing between multiple phases of the product lifecycle.

The second step involves the deployment of decision-making tools to derive alternative configurations of the production system using as input the DT models. These tools, wrapped as planning services, use AI to draw decisions on (a) how to optimally allocate the tasks among the available resources, (b) where to locate the resources for realizing optimal task plans, and (c) how to ensure collision-free resource trajectories, minimizing the stoppages of the task due to the existence of obstacles. The integration of planning services to the DT aims to bridge the gap between the physical world and its digital representation enabling production system's reconfigurable behavior [4].

The third step involves the deployment of control services for interfacing the involved hardware components, namely the robots' controller, the human–machine interfaces, the sensors, and the tools. The target is to simplify the control integration and sensor data sharing. This results in an easier distribution of derived plans to the relevant resources. During execution, the DT is dynamically updated, retrieving real-time sensor data, resources' status, and location as well as human input.

HRC Workplaces Re-Design Workflow

The proposed method aims to enable the operation of systems that employ human operators and flexible mobile robot workers. The mobility of these resources creates two-fold added value for the system: (a) the layout of the system can be autonomously reconfigured by re-directing the mobile robots in different workstations, and (b) the robots act as assistants to human operators increasing flexibility on the process execution [5].

The design process of HRC workplaces operation commonly proposed in the literature follows the sequential workflow presented in Figure 3 [29]. The result of this process is a fixed HRC layout where the tasks and robot motions are pre-determined so to comply with the safety regulations [30]. Human uncertainty is handled by isolated safety systems, leading to stoppages of production when the safety is compromised by human behavior.

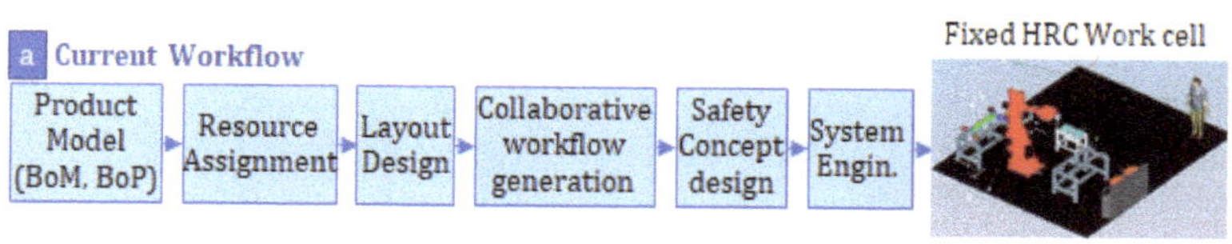

Figure 3. Current practice on the design workflow of HRC applications.

The proposed DT based re-design process aims to facilitate this dynamic system reconfiguration by transforming the design workflow, as shown in Figure 4. The DT models include the information needed for optimizing the system's layout by assigning human operators and mobile robots to perform tasks in the different workstations. During execution, a set of perception modules is deployed to monitor the execution, the environment, and human behavior through multiple sensors. These sensors capture 2D and 3D images. Consuming the relevant data structures of the DT, these data are synthesized to populate the real-time planning scene in the form of an occupancy map [31].

The decision-making logic implemented for the layout and task planning is visualized in Figure 5. An intelligent search method [1] is used to generate alternative configurations of the system consuming the 3D planning scene and the resources, tools, and task DT models. Each task model selected in the task layer is linked to the involved part model and its real-time location in the shopfloor. In the second layer, the availability of resources is assessed based on their real-time status, considering human operators (H), mobile robot workers (M.R.), and human–robot teams (H.MR.). In the third layer, based on resource capabilities including payload and available tooling, the suitability for the selected task is assessed. In the Tools layer, the fitting tooling is selected. In the last layer, the robots' motion for each alternative is calculated consuming the 3D planning scene. Finally, the alternative with the minimum execution duration is selected and sent for execution to the physical system. In cases of disturbances in runtime, the planning services are triggered to re-design the workflow.

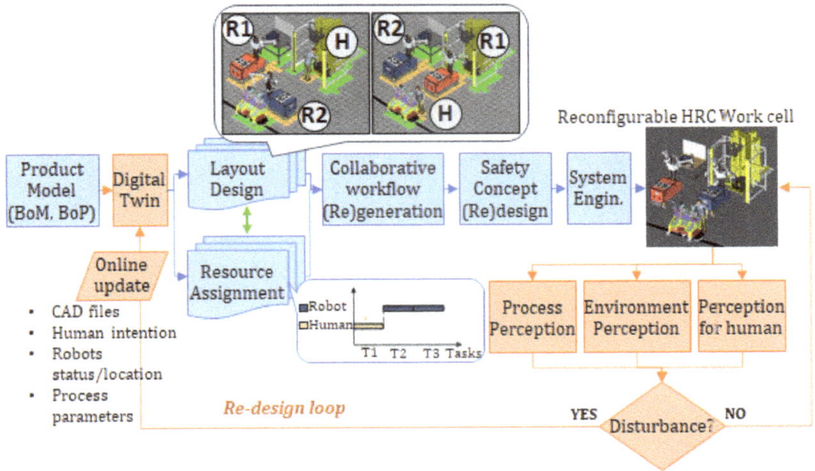

Figure 4. Proposed paradigm on the design workflow of HRC applications.

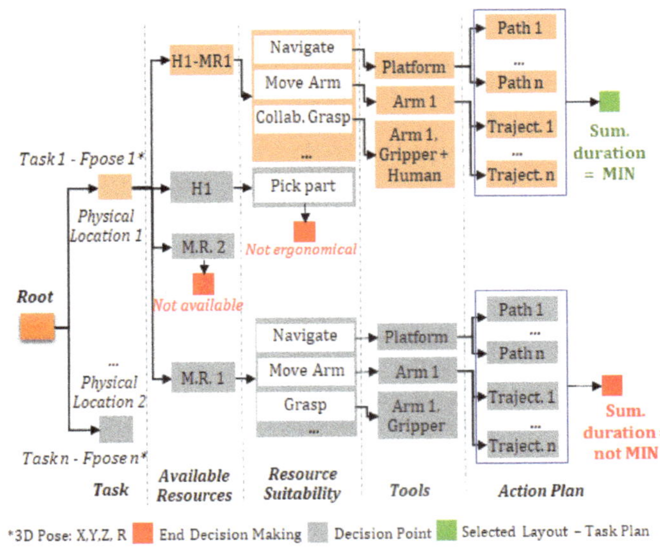

Figure 5. Decision-making logic for reconfigurable factories.

3. Implementation

The overall architecture of the Digital Twin including the components deployed for the construction of the DT is visualized in Figure 6. The integration and communication interfaces were established following the principles of ROS [31] to enable data exchange among the different modules as well as scalability of the system in terms of adding new robot resources. The system is deployed in a PC running Ubuntu 16.04 and ROS Kinetic. The involved modules are described below.

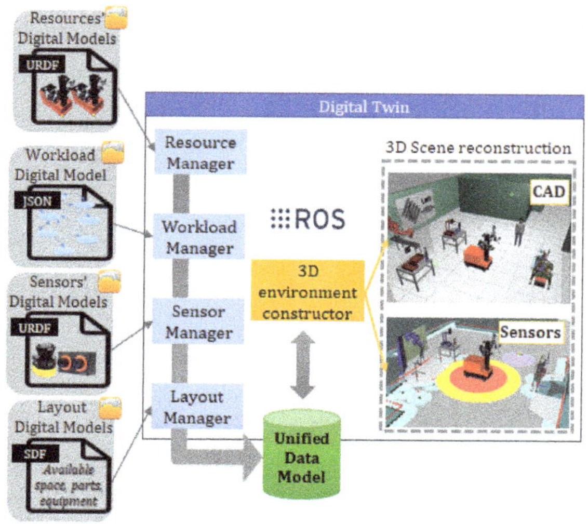

Figure 6. Digital Twin system architecture.

The Resource Manager module is responsible for storing and broadcasting the Resources' digital models. These models involve attributes such as payload, minimum velocity, location, status, kinematics transform configuration (.urdf), path (.yaml), and motion (.srdf) configuration. The Workload Manager stores the task models in a JSON format following a hierarchical modeling approach [12]. The involved sensors' configuration data such as ID, location, and Unified Robot Description Format (URDF) description are registered in the DT through the Sensor Manager. The captured data are published through dedicated topics. On top of these, the 2D and 3D data are combined and published in an additional topic so to facilitate their consumption by the planning services. The Layout Manager stores all CADs files related to static fixtures, parts, and products in .sdf format also defining the collisions, the inertia, and the mass parameters. The 3D environment constructor retrieves the locations of all parts, fixtures, sensors, and resources to construct an environment with a global world frame (using ROS Tf—library [31]). It also provides an interface with ROS services and a topic to track and update the position of all parts inside the shopfloor based on the real-time updates coming from shopfloor sensors.

The integration of the DT with the planning and control services is visualized in Figure 7. The layout and task planning modules are developed in Java. Regarding the robot motions, state-of-the-art algorithms for mobile platform path planning and robot arm motion planning are integrated in the system (gmapping, amcl, and ompl [31]). The DT provides dedicated interfaces to these modules so for the latter to have access in the 3D planning scene and exchange data. The connection with the robot controller is established through ROS drivers, while the instructions for the tasks are sent to human operators in smart devices interfaces through ROSBridge.

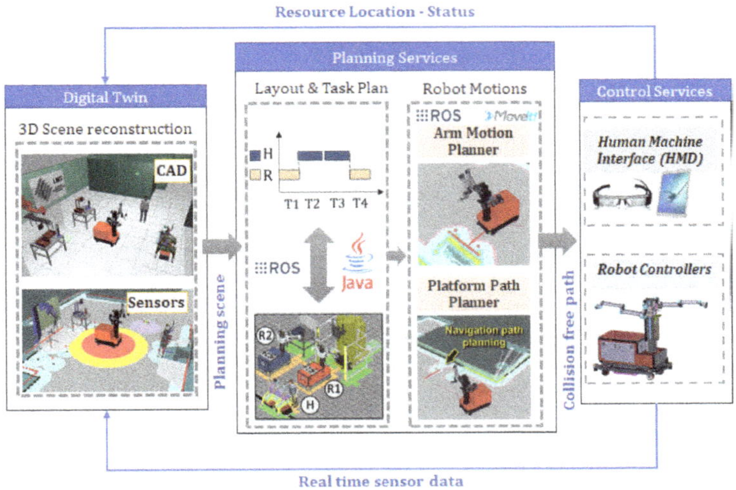

Figure 7. Digital Twin based re-design system architecture.

During execution, the DT is used for publishing the planning scene constructed through the elements described above. The component diagram presented in Figure 8 details the connection and the established interaction among the different planning and control services with the DT.

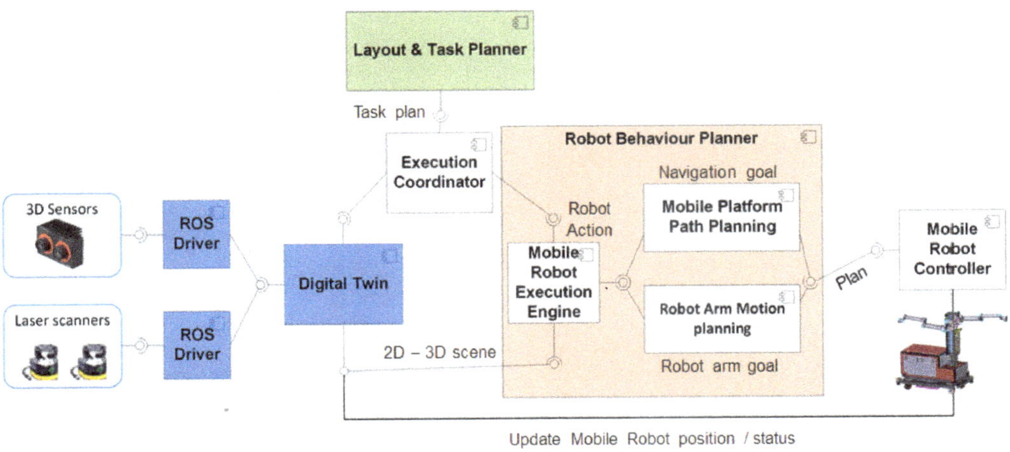

Figure 8. DT-based system component diagram.

In order to efficiently accommodate the interaction among these modules, a service-oriented approach has been implemented. Under this approach, the exchange of information among the different resources is performed using peer-to-peer services. In Figure 9, the provided sequence diagram details the communication among the different resources indicatively for the execution of a Navigation Task by a mobile robot.

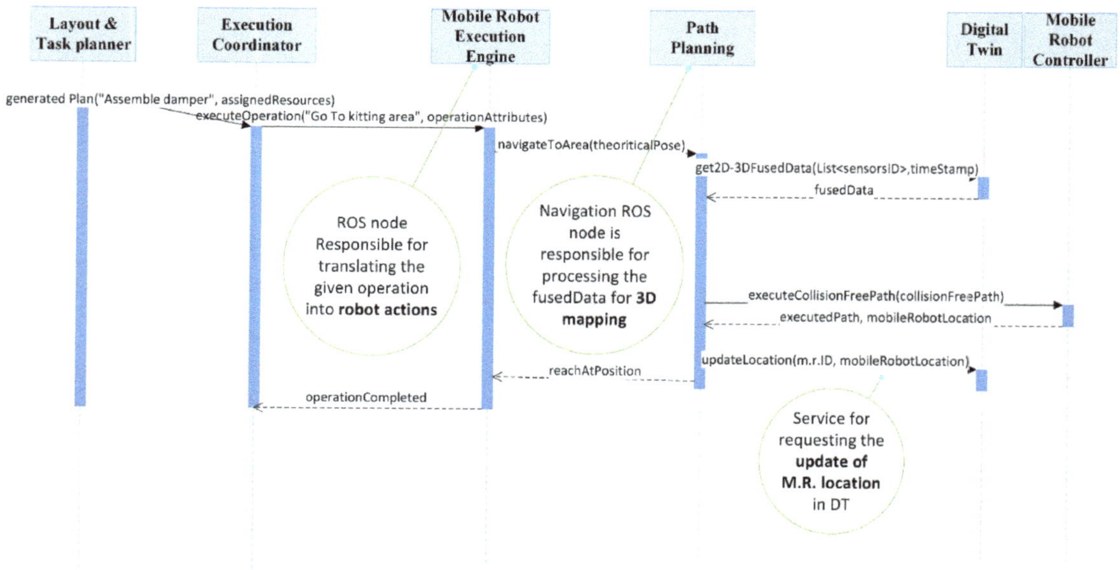

Figure 9. DT-based system sequence diagram for the execution of a Navigation Task.

4. Case Study

The section describes how the DT based re-design system has been deployed, tested, and validated in a real industrial case study. In particular, the system has been tested in an assembly line of an automotive company producing the front axle of a passenger vehicle. The current assembly line, deployed in the automotive factory, is configured in a fixed layout employing human operators, compression machines, and an automated screwing machine. The base part is loaded in an Automated Guided Vehicle (AGV) that moves along the assembly line. The process takes place in three working areas: (a) the pre-assembly working area, (b) the compression working area, and (c) the AGV working area. However, this set up raises two main challenges for the end user. (a) The first is hardware equipment limitations to accommodate the production of multiple product variants since the automated screwing machines can accommodate the screwing of only one axle model. When new models are added in production, a manual re-design and re-deployment of the system is needed. (b) The second is the strain caused in human operators, since they need to lift 480 compressed dampers of 6 kg weight in an 8 h shift. To overcome those limitations, the DT based re-design system is used to enable the production paradigm that employs mobile dual arm workers. The workers can (a) be re-located among workstations and by automatically changing tools perform operations for multiple parts and (b) undertake the heavy tasks, reducing operators' strain.

The DT-based system is used to generate alternative configurations of the system. Depending on the alternative task plan, the resources are located in different workstations, forming the initial layout of the system. Figure 10 presents three indicative system configurations for the assembly of the axle that takes place in three working areas.

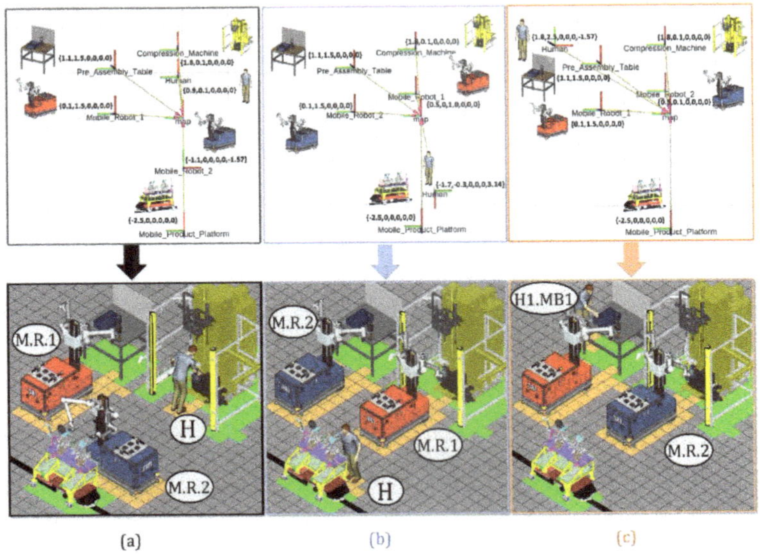

Figure 10. Indicative alternative configurations of the assembly system: (**a**) alternative 1, (**b**) alternative 2, (**c**) alternative 3.

The generated alternatives are evaluated toward minimizing the processing time and optimizing the use of resources. To validate the alternatives against those metrics, the decision-making system is integrated with a GAZEBO 3D simulator [32]. This simulator provided the values on these metrics for each configuration alternative. After all alternatives were validated, the optimal configuration was selected by the system. The selected task plan that is dispatched for execution is listed in Table 1.

Table 1. Assembly tasks allocation to resources.

Tasks	Working Area	M.R.	Human
Pre-assembly of right damper	Pre-assembly		X
Pick and place of pre-assembled damper	From pre-ass. to compress.	X	
Align damper in comp. machine	Compression		X
Remove alignment rod	Compression	X	
Pick and place nut	Compression	X	
Pick and place compressed damper	From comp. to AGV.	X	
Pick and insert cables/screws on the disk	Compression		X
Screwing damper on the disk	Compression	X	

The selected configuration of the layout has been physically deployed in the Laboratory for Manufacturing Systems and Automation (LMS) Machine Shop as visualized in Figure 11. This set up has been used for the execution of the assembly task plan as generated by the decision-making module.

Figure 11. The selected configuration physically deployed.

In this study, we consider two cases that raise the need for system re-design during execution. The first case concerns execution disturbances that call for resource behavior level re-design when the robots' motions need to be adjusted in online changes that occur in the environment. Currently, as mentioned above, AGVs are used in the factory. When a human operator gets close of one of the AGVs, the latter stops, and it only resumes its operation when the human moves away. In such a fenceless environment, these stops occur quite often during a shift, creating eventual delays in the process. This creates the need for the mobile robots to be able detect human operators, and instead of stopping, re-plan its path so to avoid them.

The second case concerns execution disturbances that call for Station Level re-design by re-organizing the remaining tasks to the involved resources. This case is of interest considering that in actual production, resources' breakdown occur during execution. When these occur, the production stops, and typically, the contribution of the maintenance department is needed to resume the execution. The target is to use the DT-based system to minimize the stoppages of production due to resources' breakdown by dynamically re-designing the tasks when these occur.

The next sub-section presents a quantification of the effect of using the DT-based system in the initial design of the system and in the occurrence of the two type of disturbances.

Results

Regarding the design of the production system, Discrete Event Simulation (DES) has been used for calculating the performance of the DT-based generated design assembly paradigm against current practice. Two different DES models were deployed in WITNESS Discrete Event Simulation Software, version 23 [33]. The first model represented the current assembly line structure, modeling the execution time for each workstation. The tasks involved in each workstation were executed by human operators and by a compression machine. The execution time for these tasks was modeled based on relevant data provided by the automotive end user. The second model simulated the new set up of the production system that employs mobile robots to support human operators. The structure of the second model was designed based on the outcome of the DT-based system. This model as well incorporated the execution time of the tasks performed by the mobile robots, the human operators, and the compression machine. The execution time for human operators and the compression machine was modeled using the same source of information as in the first model. For modeling the time required for the mobile robots to execute their assigned tasks, the execution of the tasks by the robots was simulated in GAZEBO 3D simulator recording their duration.

The simulation time for both models was one year. The production throughput and the utilization of resources were recorded for both models during this simulated year of production. The results, shown in Table 2, indicate a considerable improvement in the production throughput and utilization of operators' availability by facilitating the reconfigurable HRC system.

Table 2. Comparison of the two systems—designed system performance.

KPI	Current	DT Based Design	Increase
Annual production throughput—No. of Parts	315,353	394,997	25%
Time percentage allocated to performing assembly (%)	34	49	30%

Regarding the re-design of the system, below, we present the responsiveness of the system in two cases of execution disturbances against current industrial practice. Figure 12 visualizes the 3D planning scene reconstructed by the DT based on the physical world real-time status. This 3D planning scene is generated by the combination of data captured by two laser scanners placed on the mobile platform and one stereo camera placed on the robot's torso. By combining this information, the human presence and state is identified and included in the planning scene.

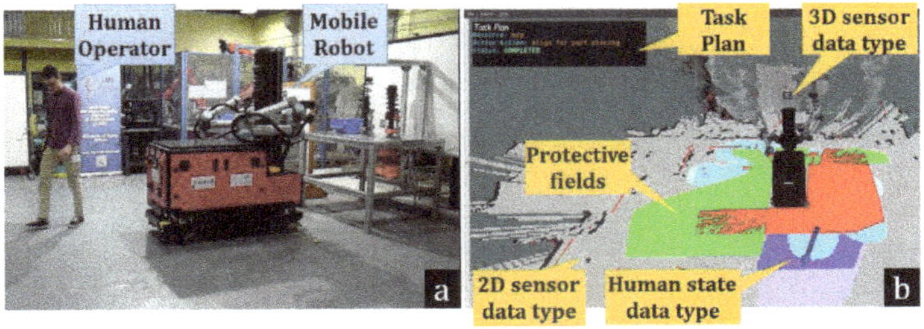

Figure 12. Automotive pilot cell (**a**) real world, (**b**) 3D planning scene.

Case 1: The operator enters the common workspace, and his path intersects the mobile robot's planned trajectory for 10 s. In current industrial practice, the mobile robot would stop, waiting for the human to move away and needing 1 s to resume its path. In the same case, the DT-based system will detect his NO INTERACTION intention and add human presence and state information in the planning scene. The mobile robot, instead of stopping, will re-plan its path in around 2 s, as shown in Figure 13. To quantify this effect, in an 8 h shift, we consider that the human will block the robot path 200 times. As shown in Table 3, the latter practice may reduce by 80% the overall delay caused by human presence.

Figure 13. Human NO INTERACTION intention detected.

Table 3. Comparison of the two systems—responsiveness to disturbances.

Time to Re-Design the Execution/Shift	Current	DT Based	Decrease
Case 1—Robot behavior re-design	36 (min)	7 (min)	80%
Case 2—Station level re-design	58 (min)	17.5 (min)	69%

Case 2: Robot arm fails to grasp the damper. In a realistic set up, the robot may fail to execute a task due to occlusions in the cameras' field of view, lighting conditions, etc. Through experimentation, we consider that in an 8 h shift, the robot fails 50 times. In current practice, the production manager would manually re-start the robot and correct its position, needing up to 70 s. Through our system, the operator himself can easily resume the execution by manually guiding the arm in the grasping position, as shown in Figure 14, needing up to 21 s. Table 3 shows the time saved by using the DT model functions.

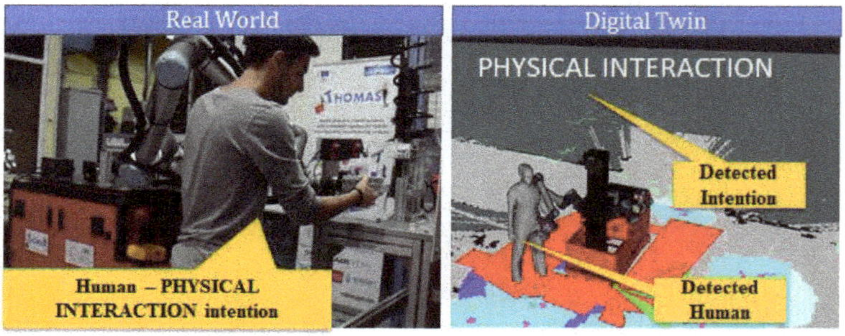

Figure 14. Human PHYSICAL INTERACTION intention detected.

5. Conclusions

This work discusses the use of DTs for designing and redesigning flexible production systems. These systems employ mobile dual arm robots that move across the factory undertaking multiple tasks, assisting humans. Exploiting this hardware ability, the DT based re-design system generates optimal configurations in terms of layout and task plans. The solution allows for online reconfiguration of the system by (a) dynamically re-assigning the tasks when unexpected events occur, (b) real time adjusting robots' behavior so to ensure collision-free trajectories generation in unstructured environments. These are achieved reasoning on top of the sensor-based real-time scene reconstruction that is provided by the Digital Twin by synthesizing multiple sensor data.

The discussed DT-based system has been deployed, tested, and validated in a case study from the automotive sector. In particularly, mobile robot workers are introduced to assist human operators in the assembly of a vehicle's front axle, which is performed fully manually in current practice. Discrete Event Simulation has been used to validate the performance of the suggested hybrid production system against current practice. The results show that by introducing mobile robot resources in the system, the production throughput as well as the resources' utilization can be increased. To validate the effect of the DT-based system that enables this hybrid production paradigm, its effect in two cases of needed reconfiguration has been analyzed and quantified. This quantification proves that the suggested system enables the faster adaptability of the system when unexpected events occur against current practice, minimizing the stoppages of production

Future work will aim to a higher Technology Readiness Level (TRL) of the application so that it may be deployed in an industrial environment. Furthermore, the target is to enhance the digital models and decision-making tools so to incorporate the integration of dynamic risk assessment tools ensuring compliance to safety regulations. In addition, the DT-based system may be extended so to include the modeling and planning of the intra factory logistics operations required for supporting the assembly as well as for respecting the desired by the end user flowtime for the products.

Author Contributions: Conceptualization, S.M., G.M. and N.K.; methodology, N.K. and C.G.; software, C.G., K.L., A.C.B. and S.A.; validation, S.A. and P.B.; formal analysis, N.K. and S.M.; investigation, N.K., C.G. and K.L.; resources, S.M.; data curation, A.C.B.; writing—original draft preparation, N.K. and S.A.; writing—review and editing, N.K. and S.M.; visualization, C.G. and P.B.; supervision, N.K. and S.M.; project administration, N.K., G.M. and S.M.; funding acquisition, G.M. and S.M. All authors have read and agreed to the published version of the manuscript.

Funding: This research was funded by European Union's Horizon 2020 research and innovation programme under grant agreement No. 723616 (www.thomas-project.eu/, accessed on 13 May 2021) and 825196 (https://trinityrobotics.eu/, accessed on 13 May 2021).

Institutional Review Board Statement: Not applicable.

Informed Consent Statement: Not applicable.

Data Availability Statement: Not applicable.

Conflicts of Interest: The authors declare no conflict of interest.

References

1. Chryssolouris, G. *Manufacturing Systems: Theory and Practice*, 2nd ed.; Springer: New York, NY, USA, 2006.
2. Michalos, G.; Kousi, N.; Makris, S.; Chryssolouris, G. Performance assessment of production systems with mobile robots. *Procedia CIRP* **2016**, *41*, 195–200. [CrossRef]
3. Stark, R.; Kind, S.; Neumeyer, S. Innovations in digital modelling for next generation manufacturing system design. *CIRP Ann.* **2017**, *66*, 169–172. [CrossRef]
4. Makris, S. *Cooperating Robots for Flexible Manufacturing*; Springer: Berlin/Heidelberg, Germany, 2020.
5. Kousi, N.; Michalos, G.; Aivaliots, S.; Makris, S. An outlook on future assembly systems introducing robotic mobile dual arm workers. *Procedia CIRP* **2018**, *72*, 33–38. [CrossRef]
6. Erkoyuncu, J.A.; del Amo, I.F.; Ariansyah, D.; Bulka, D.; Roy, R. A design framework for adaptive digital twins. *CIRP Ann.* **2020**, *69*, 145–148. [CrossRef]
7. Kousi, N.; Koukas, S.; Michalos, G.; Makris, S. Scheduling of smart intra—Factory material supply operations using mobile robots. *Int. J. Prod. Res.* **2019**, *57*, 801–814. [CrossRef]
8. Tomiyama, T.; Lutters, E.; Stark, R.; Abramovici, M. Development capabilities for smart products. *CIRP Ann.* **2019**, *68*, 727–750. [CrossRef]
9. Schleich, B.; Anwer, N.; Mathieu, L.; Wartzack, S. Shaping the digital twin for design and production engineering. *CIRP Ann.* **2017**, *66*, 141–144. [CrossRef]
10. David, J.; Martikkala, A.; Lobov, A.; Lanz, M. A Unified Ontology Namespace for Enterprise Integration—A Digital Twin Case Study, Instrumentation Engineering. In Proceedings of the Instrumentation Engineering, Electronics and Telecommunications, Izhevsk, Russia, 24–26 November 2019; pp. 13–22.
11. Tao, F.; Zhang, M.; Liu, Y.; Nee, A.Y.C. Digital twin driven prognostics and health management for complex equipment. *CIRP Ann.* **2018**, *67*, 169–172. [CrossRef]
12. Tsarouchi, P.; Spiliotopoulos, J.; Michalos, G.; Koukas, S.; Athanasatos, A.; Makris, S.; Chryssolouris, G. A Decision Making Framework for Human Robot Collaborative Workplace Generation. *Procedia CIRP* **2016**, *44*, 228–232. [CrossRef]
13. Erdős, G.; Paniti, I.; Tipary, B. Transformation of robotic workcells to digital twins. *CIRP Ann.* **2020**, *69*, 149–152. [CrossRef]
14. Kousi, N.; Gkournelos, C.; Aivaliotis, S.; Giannoulis, C.; Michalos, G.; Makris, S. Digital twin for adaptation of robots' behavior in flexible robotic assembly lines. *Procedia Manuf.* **2019**, *28*, 121–126. [CrossRef]
15. Barricelli, B.R.; Casiraghi, E.; Fogli, D. A Survey on Digital Twin: Definitions, Characteristics, Applications, and Design Implications. *IEEE Access* **2019**, *7*, 167653–167671. [CrossRef]
16. Florescu, A.; Barabas, S.A. Modeling and Simulation of a Flexible Manufacturing System—A Basic Component of Industry 4.0. *Appl. Sci.* **2020**, *10*, 8300. [CrossRef]
17. Florescu, A.; Barabas, S.A. Simulation Tool for assessing the performance of Flexible Manufacturing Systems. In *IOP Conference Series: Materials Science and Engineering, Proceedings of the 2nd International Conference on Manufacturing Technologies (ICMT 2018), Orlando, FL, USA, 19–21 January 2018*; IOP Publishing: Bristol, UK, 2018; Volume 398, p. 012023.
18. He, B.; Bai, K.J. Digital twin-based sustainable intelligent manufacturing: A review. *Adv. Manuf.* **2021**, *9*, 1–21. [CrossRef]
19. Shao, G.; Helu, M. Framework for a digital twin in manufacturing: Scope and requirements. *Manuf. Lett.* **2020**, *24*, 105–107. [CrossRef] [PubMed]
20. Resman, M.; Protner, J.; Simic, M.; Herakovic, N. A Five-Step Approach to Planning Data-Driven Digital Twins for Discrete Manufacturing Systems. *Appl. Sci.* **2021**, *11*, 3639. [CrossRef]

21. Schütze, A.; Helwig, N.; Schneider, T. Sensors 4.0–smart sensors and measurement technology enable Industry 4.0. *J. Sens. Sens. Syst.* **2018**, *7*, 359–371. [CrossRef]
22. Tao, F.; Zhang, H.; Liu, A.; Nee, A.Y. Digital twin in industry: State-of-the-art. *IEEE Trans. Ind. Inform.* **2018**, *15*, 2405–2415. [CrossRef]
23. Novak, D.; Riener, R. A survey of sensor fusion methods in wearable robotics. *Robot. Auton. Syst.* **2015**, *73*, 155–170. [CrossRef]
24. Nagla, K.S.; Uddin, M.; Singh, D. Multisensor Data Fusion and Integration for Mobile Robots: A review. *IAES Int. J. Robot. Autom.* **2014**, *3*, 131. [CrossRef]
25. Santoso, F.; Garratt, M.A.; Anavatti, S.G. Visual—inertial navigation systems for aerial robotics: Sensor fusion and technology. *IEEE Trans. Autom. Sci. Eng.* **2016**, *14*, 260–275. [CrossRef]
26. Bilberg, A.; Malik, A.A. Digital twin driven human–robot collaborative assembly. *CIRP Ann.* **2019**, *68*, 499–502. [CrossRef]
27. Quigley, M.; Conley, K.; Gerkey, B.; Faust, J.; Foote, T.; Leibs, J.; Wheeler, R.; Ng, A.Y. ROS: An open-source Robot Operating System. In Proceedings of the ICRA Workshop on Open Source Software, Anchorage, AK, USA, 12–17 May 2009; IEEE: New York, NY, USA, 2009.
28. Stark, R.; Fresemann, C.; Lindow, K. Development and operation of Digital Twins for technical systems and services. *CIRP Ann.* **2019**, *68*, 129–132. [CrossRef]
29. Michalos, G.; Kousi, N.; Karagianis, P.; Gkournelos, C.; Dimoulas, D.; Koukas, S.; Mparis, K.; Papavasiliou, A.; Makris, S. Seamless human robot collaborative assembly—An automotive case study. *Mechatronics* **2018**, *55*, 194–211. [CrossRef]
30. Michalos, G.; Makris, S.; Tsarouchi, P.; Guasch, T.; Kontovrakis, D.; Chryssolouris, G. Design Considerations for Safe Human–Robot Collaborative Workplaces. *Procedia CIRP* **2015**, *37*, 248–253. [CrossRef]
31. Pyo, Y.; Cho, H.; Jung, L.; Lim, D. *ROS Robot Programming*; ROBOTIS: Seoul, Korea, 2017.
32. GAZEBO Simulator. Available online: http://gazebosim.org/ (accessed on 13 May 2021).
33. Lanner Witness Simulation. Available online: https://www.lanner.com/en-us/technology/witness-simulation-software.html (accessed on 13 May 2021).

applied sciences

MDPI

Article

Manufacturing Execution System Integration through the Standardization of a Common Service Model for Cyber-Physical Production Systems

Richárd Beregi *, Gianfranco Pedone, Borbála Háy and József Váncza

Centre of Excellence in Production Informatics and Control, Institute for Computer Science and Control, Kende u. 13-17, 1111 Budapest, Hungary; gianfranco.pedone@sztaki.hu (G.P.); borbala.hay@sztaki.hu (B.H.); jozsef.vancza@sztaki.hu (J.V.)

* Correspondence: richard.beregi@sztaki.hu

Abstract: Digital transformation and artificial intelligence are creating an opportunity for innovation across all levels of industry and are transforming the world of work by enabling factories to embrace cutting edge Information Technologies (ITs) into their manufacturing processes. Manufacturing Execution Systems (MESs) are abandoning their traditional role of legacy executing middle-ware for embracing the much wider vision of functional interoperability enablers among autonomous, distributed, and collaborative Cyber-Physical Production System (CPPS). In this paper, we propose a basic methodology for universally modeling, digitalizing, and integrating services offered by a variety of isolated workcells into a single, standardized, and augmented production system. The result is a reliable, reconfigurable, and interoperable manufacturing architecture, which privileges Open Platform Communications Unified Architecture (OPC UA) and its rich possibilities for information modeling at a higher level of the common service interoperability, along with Message Queuing Telemetry Transport (MQTT) lightweight protocols at lower levels of data exchange. The proposed MES architecture has been demonstrated and validated in several use-cases at a research manufacturing laboratory of excellence for industrial testbeds.

Keywords: interoperability; industry 4.0; manufacturing execution system; cyber-physical production system; OPC UA

Citation: Beregi, R.; Pedone, G.; Háy, B.; Váncza, J. Manufacturing Execution System Integration through the Standardization of a Common Service Model for Cyber-Physical Production Systems. *Appl. Sci.* **2021**, *11*, 7581. https://doi.org/10.3390/app11167581

Academic Editor: Emanuele Carpanzano

Received: 16 July 2021
Accepted: 16 August 2021
Published: 18 August 2021

Publisher's Note: MDPI stays neutral with regard to jurisdictional claims in published maps and institutional affiliations.

1. Introduction

There is beauty in how complex the human body is, yet seemingly how naturally and effortlessly it operates day-to-day. For the untrained eyes, a manufacturing plant can seem such a complicated, yet well-functioning entity. Both have a system which connects faculties of sensing, decision-making, and acting, which is essential for their proper functioning [1]. For the human body, this is obviously the nervous system. For a production facility, this is the *Manufacturing Execution System*, or MES in short.

Many definitions and approaches of MES appeared, yet somehow all of them tried to modularize and incorporate every aspect of manufacturing and to combine them into one monolithic system [2]. Thus far, there has not been a minimalist, core concept of MES. One which only has the most basic, *core functionalities* to execute a manufacturing plan. The aim was to take out every intelligent and decision-making aspects of MES to create a *nervous system for manufacturing*, which connects both low-level resources and their operations together, in addition to linking them to the high-level orchestrators, like a scheduler or an Enterprise Resource Planner (ERP). This approach operates with the expectation that both low- and high-level components have built-in intelligence to enable smart production through the digitalization of their environment. The basis for this expectation is the raising phenomena of Cyber-Physical Systems (CPSs) in the manufacturing shop-floor and the spread of generally accepted distributed decision-making supporting the *Cyber-Physical Production System* (CPPS) concept [3,4].

Manufacturing digitalization prevails more and more, and industrialized countries now have their own initiatives to support it [5]. Industry 4.0 (I4.0) is all about integration and digitalization [6,7], and, given the crucial role of data and Industrial Internet of Things (IIoT) in its technological scenarios, there is a common view that integrated MES are still crucial, although moving toward the broader Manufacturing Operations Management (MOM) and service-oriented approach [8]. Next to this, robotics is expanding in magnitude around the developed world, also in executing manufacturing processes, and is expected to rise from a 15 billion US dollar sector now to 67 billion by 2025 [9].

Overall, I4.0 is demanding a gradual but continuous evolution towards interoperable standards and systems, distributed and self-learning MES services [10]. In this view, modular (not monolithic) MES, with functions becoming services, acquires a central role. Customers in manufacturing require support to multi-plant enterprise orchestration, easy scalability, optimal interplay with modern technologies, and process innovation support through the integration with additive manufacturing, robotics, and Plant Lifecycle Management (PLM) systems. In order to digitally transform industrial automation, significant efforts have to be dedicated to the isolated systems [11].

As I4.0 grows, MES need to grow alongside it [12]. Global networking of production resources and processes, as well as the use of globally interconnected applications, crucially require the adoption of unified standards. *Standardization* in manufacturing architectures evolved from ISA95 to Manufacturing Enterprise Solutions Association International (MESA) [13] in the past decades and, more recently, to service-oriented industrial reference architectures, such as RAMI 4.0 (Reference Architecture Model for Industry 4.0) [14] and IIRA (Industrial Internet Reference Architecture) [15]. All these have proved their applicability also to MES and CPPS, with the intention to organize, define, and integrate automated functions (services) that can be connected to the *core*. These two major standardization proposals are *de-facto* considered reference guidelines for conceiving future business organizations, even though a final consensus or agreement in the scientific and industrial communities has not been reached yet. An introduction to the standardization efforts is reported in Section 2.4, whereas a more comprehensive, cloud- and service-oriented presentation, and comparison of both reference architectures can be found in [16].

This evolutionary process is not straightforward, and an analysis reported also in [10] investigated the evolution of *distributed intelligence architectures* and their expected path of development, what many today identify with Multi-Agent Systems (MAS) [17]. While IIoT is a growth factor for MES in some industries, the availability of newer technologies and approaches can hinder MES deployments. Companies are trying to reduce their future risks and understand future directions in the plethora of I4.0 standards [18]. They are willing to adopt a technology which will be part of the new standards, which is easy to integrate, use, and expand, as well as which can flexibly accommodate new technologies and applications as they are developed. Research work outlined in this paper aimed ultimately at identifying the minimum set of service concepts and technologies that every CPPS should commonly understand and leverage in their standardized service provision.

In line with the argumentation reported in [19], I4.0 is requiring also a new style of MES for several reasons. Firstly, modern consumers are demanding customization. I4.0 offers an unprecedented opportunity for this, but companies need a plant floor software that is ready to adapt to the underlying technologies. Secondly, I4.0 is a concept revolving around CPSs, which is about the merging of virtual and real worlds and the creation of links between information and operations technologies. Thirdly, reference architectures are not always easily nor efficiently applicable in small to mid sized development projects due to their complexity and design requirements.

However, all the above issues cannot be resolved overnight and MES needs to be upgraded alongside the challenges and opportunities of I4.0. In particular, the shop-floor has to be optimized through decentralized management, where autonomous agents may be seen as a marketplace with demand (the products being built) and supply (the equipment). Future MES needs to accommodate connectivity, mobile, cloud, and advanced service-oriented computing, as well as human collaboration in order to face the challenges of a new industrialization era [19]. Supporting Human–Robot Collaboration (HRC) in a multi-agent setting is a current issue of special relevance [20]. Furthermore, MES needs context-resolution capabilities in order to enable components of a CPPS to act autonomously and release their own intelligence even in a decentralized setting. In this sense, an MES is an orchestrator of complex production processes, typically over multiple layers of abstraction. Finally, IIoT sensory activity and CPPS communications create new data flows and their vertical integration is crucial to ensure an effective response in the enterprise.

1.1. Main Contributions

In the face of new requirements, typical MES functions still remain necessary in the I4.0 era as well, even though in a renewed architecture. Just as the traditional automation pyramid is evolving, the MES should also be changed accordingly, towards providing services. Figure 1 shows how the development process of the MESA automation pyramid is extended with the current proposal in this paper.

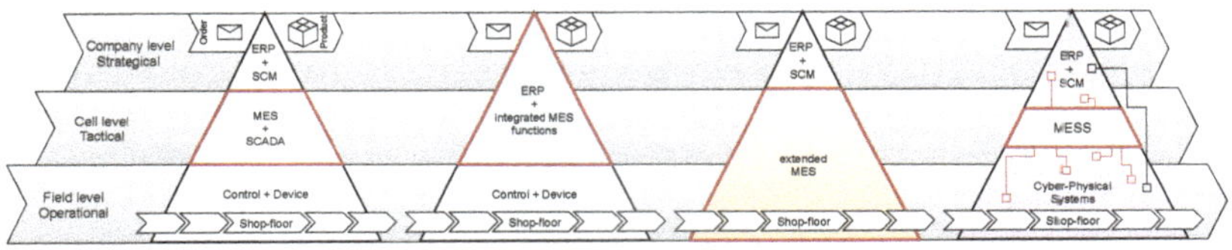

Figure 1. Different functional positioning of Manufacturing Execution System within a company.

Following the above considerations, research presented in this paper targeted the design and realization of a novel concept of *Manufacturing Execution System as a Service* (MESS in the following). This is a new attempt to model, integrate, and orchestrate essential CPS services that make up a modern digitized manufacturing facility. The architecture of MESS facilitates a novel, generic, and simple way for the collaboration of distributed, autonomous manufacturing entities. The focus of the paper is on the core functionality of MESS and the standardized integration of CPS services.

Major points of this research work can be summarized as follows:

(i) Proposition of a simplified CPS conceptual and information model, whose elements represent the basis for the definition of a common vocabulary in service integration and interoperability semantics (see Section 3.2);
(ii) Realization of such service integration model as the core of a new, minimalistic concept of MES, which is based on as few assumptions as possible. It privileges standardized industrial technology service and data publishing via Open Platform Communications Unified Architecture (OPC UA) and Message Queuing Telemetry Transport (MQTT) (see Section 3.3);
(iii) Exploitation of such universal service integration model to a variety of heterogeneous manufacturing cells through by means of a so-called Production Administration Shell (PAS) (see Section 4.3);
(iv) Demonstration of the presented approach in a research lab for manufacturing. This is a learning factory and an open ecosystem which offers students, researchers, as well as industrial partners the opportunity to perform research on CPPS related topics. An ideal place to study the challenges and to understand the benefits of elevating the CPS to a mature level of interoperability in a production context (see Section 5).

1.2. Paper Outline

After an introduction of I4.0 industrial requirements and the importance to migrate from traditional MES to CPPS-based interoperable environments, motivation and contributions are presented in short. Section 2 gives a short overview of related works and aims at specifying the goals and requirements of the system targeted in this research. In Section 3, the basic concepts underpinning the methodology are enumerated, and the methodological approach is presented in detail, highlighting the fundamentals of the service integration model, its OPC UA counterpart, as well as the overall MESS system architecture. Section 4 is centered around the realization of the MESS; meanwhile, Section 5 details the application and demonstration of this work in a pilot case study. Discussion and conclusions close the paper.

2. Review of Literature and Standards

The aim of this review is to find the core manufacturing and IT functionalities of a MES that can meet and serve the requirements of CPPS. An overview of recent MES solutions has been performed with a special emphasis on those which could support distributed smart manufacturing. Standardization efforts are reviewed both in regard to expected manufacturing functionalities and applied IT.

2.1. Cyber-Physical Production Systems

An overview of MES's specific role in I4.0 is presented in [21], where it is seen as the digital twin of future factories with an ability to connect real(-time) manufacturing processes with their digital images. I4.0 is the turning point which marks the end of conventional centralized applications: these new developments are scanned in [22], along with the analysis of so-called enabling technologies. Authors in [23] provide an example of basic resource virtualization for the development of CPPS, as a viable process for companies to create digital twins by specifying the digital twin hierarchy, the information to be modeled, and the modeling methods. Anyway, the referred work is not oriented to a possible generalization of the CPPS service interoperability. In [24], a framework for the classification of CPPS developments provided by the scientific society is introduced. As evidenced also in [25], connectivity represents a prior challenge in establishing CPPS on the basis of heterogeneous IT-software environments. An attempt to realize an OPC-UA based service-oriented communication model for CPPS can be found in [26], whose core is centered around the concept of a service bus. In [27], the authors focus on the automated creation of the CPPS digital twin in a body-in-white production system. The outcome of [28] elaborates on the advantages of migrating legacy IT systems for manufacturing operations to a microservice architecture, as an important step towards platform-based ecosystems. Authors in [29] present an Open CPPS automation architecture with the intention to enable vertical integration to become a reality through a CPPS architecture and IEC-61499 standard. Research work in [30] proposes an open architecture for collaborative edge and cloud processing mode, which promises fast operation and upgrading of cloud manufacturing systems through smart monitoring-analysis-planning-execution in a closed loop. An example of architecture that implements CPPS in the cloud with distributed control and the promise to provide fault tolerance are proposed in [31], but authors primarily focus on load balancing and clustering. Research work presented in [32] refers to a CPPS integrated architecture with OPC UA server for 5G-based Smart Manufacturing, but the approach is centralized. Authors in [33] propose a conceptual model of structural and behavioral elements in CPPS, but the attempt to address the gaps of semantic description tools to be included in behavioral elements and in conceptual description remains more at a theoretical stage. A common ontology model in web-based digital twin modeling and remote control of CPPS is proposed in [34]. Finally, an attempt of dynamic reconfiguration of service-oriented resources in CPPS via process-independent approach and multiple criteria for resources and operations can be found in [35].

2.2. Distributed Intelligence Architectures in Manufacturing

Distributed artificial intelligence (DAI) has emerged as a branch of Artificial Intelligence (AI) over thirty years ago as a powerful paradigm for representing and solving complex problems. The growth of the field has been spurred by the advances in distributed computing environments and the widespread connectivity. Representing a confluence of ideas from several disciplines, DAI became an independent research discipline with two main sub-fields: distributed problem solving and multi-agent systems.

Distributed problem solving deals with the issues related to solving a problem by dividing it among a number of cooperative problem solvers which share the computational burden and knowledge of their partial solutions. *Multi-agent systems*, on the other hand, are concerned with the behavior of loosely coupled autonomous problem solvers, or agents that work together to solve a problem beyond their individual capabilities. Since the DAI field is closely associated with the notion of agents, it is often referred to as multi-agent system research in literature. An important characteristic of DAI is that it combines computational resources of a group of agents such that the group intelligence is more than the sum of the individual agents' capabilities.

Agents made a real break-through two decades ago or so when the emphasis in mainstream AI research shifted from goal-seeking, logic-centered to utility-oriented, rational behavior—from single to multiple, collaborative problem-solving entities. It only mattered that an agent did the "right thing", even with bounded computational resources [36]. Accordingly, after making observations, it changed its environment by its actions in a way which served best its own interest. The impacts of actions were qualified, and not the mechanisms that generated them. This pragmatic, albeit theoretically founded concept of AI intensified further research in machine learning (when improving the performance of an agent-based on its accumulated experience), in multi-agent systems (when the environment is populated by other agents, too), and also in robotics (when integrating collaborative human and machine agents) [20].

The take-up of agent technologies was accelerated by the development of powerful multi-agent design methodologies, multi-agent simulation as well as programming environments. The agent-based paradigm of computing was early welcome in manufacturing systems control as well [37], with the promise of such much-sought properties like autonomy, cooperation, responsiveness, redundancy, distributedness, and openness [17]. Indeed, it provided an alternative to the traditional, by design centralized, hierarchical, and rigid planning and control schemes. Its current status, achieved advancements, applicability and future outlook are continuously evolving, due to the fact that today's manufacturing systems are becoming increasingly complex, dynamic, and connected [38].

However, even though MAS as a paradigm seems to be suitable to address many of the current requirements imposed by cyber-physical production systems, it still has a long and difficult path for a wider acceptance in industry [39]. Manufacturing services must comply with strong requirements in terms of reliability, robustness, and latency, and solution providers are expected to ensure that agents will operate within certain boundaries of the production, and mitigate unattended behaviors during the execution of manufacturing activities. The so-called Holonic Manufacturing Systems (HMS) addressed these issues in some way, by introducing autonomous, collaborative entities (named holons) to represent the main aspects of manufacturing in terms of product, resource, order, and staff holons. The Product–Resource–Order–Staff Reference Architecture for HMS (PROSA) gained its name from this concept [40]. Holons, as wrappers, hid unnecessary details of the main entities and could realize a spectrum of control schemes, from strictly hierarchical via heterarchical to fully distributed ones [41]. Later, the so-called Activity Resource Type Instance Reference Architecture (ARTI) was born, which combined the strength of the holonic concept and the ideas of typing and instantiating from the information and computer sciences.

On the basis of the PROSA reference architecture, a holonic manufacturing execution system was implemented to support networked production [42]. Here, smart objects representing products found their way of realization in a continuous interplay with intelligent resources. This basically self-organized design opened new research frontiers for investigating issues of distributed intelligence, collaboration, and trust in the context of manufacturing, but in terms of scalability, standardization, and performance guarantees it could not match industrial expectations. In the context of networked production, recently, Ref. [43] suggested a mutualistic resource sharing approach, where the matching between resource offers and requests were made possible by an intermediate *platform*. It is important to note that all decisions were made locally, by the participating autonomous entities. The platform provided only means for information exchange, logical matching of bids demand and supply, and for evaluating and keeping track of the trustworthiness of the partners. Agent-based simulation was used to compare different information exchange mechanisms with respect to utilization rate, service level, and communication load. The findings were applied in the design of an industry-motivated crowdsourced manufacturing platform as well [44].

As an alternative response to the concerns of distributed control in manufacturing, recently a Manufacturing Agent Accountability Framework has been proposed, which is a dynamic authorization framework that defines and enforces boundaries in which agents are freely permitted to exploit their intelligence to reach individual and collective objectives [45].

2.3. MES from the Manufacturing Viewpoint

A MES can be defined by its functionalities according to editor from MPDV Mikrolab (DE) in [2]. Many different MES views and definitions are present in parallel; however, there are some very prominent international or nationally de-facto well-accepted industrial foundations and standardization associations, whose role is outstanding in the MES scene. The biggest three are the Manufacturing Enterprise Solutions Association International (MESA, USA), the National Institute of Standards and Technology (NIST, USA), and the Zentralverband Elektrotechnik- und Elektronikindustrie eV (ZVEI), which are the most well-known in industry.

MESA's MESA-11 model stood the test of time. It was published in 1997, yet, over the years, it obtained many extensions (integrated MES, collaborative MES); however, the eleven *core manufacturing functionalities* stayed the same. They defined MES as an information delivering component from order launch to finished goods, which enables the optimization of production activities and based on the data acquisition functionality it (i.) guides, (ii.) initiates, (iii.) responds to, and (iv.) reports on plant activities as they occur in nearly real time [46]. The numbered activities highlight the crucial functionality expected from a MES.

According to the NIST, MES is a system that uses network computing to automate production control and process automation by downloading routings and schedules, furthermore, uploading production reports. Simply put, the MES bridges the information gap between the business and the shop-floor (control) layer [47]. Here, again, this bridge role reappears and the demand to operate automatically. This is a less specific functionality requirement list; however, NIST released the SIMA (Systems Integration of Manufacturing Applications) Reference Architecture, which contains the accurate manufacturing activities necessary to realize a MES [48].

The ZVEI's VDI (Verein Deutscher Ingenieure) Guideline 5600 categorizes MES in the area of discrete manufacturing as a "process-oriented manufacturing management system" and as the "comprehensive driving force for the organization and execution of the production process" [49]. SEMATECH's (Semiconductor Manufacturing Technology Consortium) CIM (Computer Integrated Manufacturing) Framework defines Manufacturing Information and Execution Systems (MIES), which focus primarily on operations in a factory: resource integration and workpiece management and movement [50].

The ideal MES according to [2], from a functional point of view, consists of guarantees of communication with corporate management applications and with the manufacturing environment, in addition to available function groups that can be activated and used depending on the requirements. These three uncovered function groups are (i.) production, (ii.) quality, and (iii.) personnel, from which the production is relevant in this case and its seven specified sub functionalities. The previously presented MES concepts and their functionalities are listed in Table 1.

Table 1. List of major functional requirements of the Manufacturing Execution System.

Framework, Association/Company (Headquarters, Year of Publication)	Proposed Functionalities
SIMA, NIST (USA, 1987)	production data collection production unit and resource tracking production unit dispatching operation sequence development detailed schedule development data analysis document management
MESA-11, MESA (USA, 1997)	data collection/acquisition operation/detail scheduling product tracking production unit dispatching resource allocation and status process management labor management quality management maintenance management performance analysis document control
MIES, SEMATECH (USA, 1999)	material movement machine control advanced process control factory labor process specification management schedule management factory services factory management and operations
MES, ZVEI (DE, 2007)	data acquisition and processing operating resources management material management personnel management interface management information management detailed planning and detailed scheduling control quality management performance analysis
ideal MES, MPDV Mikrolab (DE, 2007)	production data acquisition machine data collection control station planning table tool and resource management transmission of machine settings material and production logistics

2.4. *MES from IT Standpoint*

With the rise of the fourth industrial revolution, many attempts to design a general IT modeling and implementation standard have been made. None of the proposed architectures have gained a unique, well-accepted reference role (mainly due to their application-domain specific nature); however, they have become a sort of de-facto expected requirement in the industrial panorama. Hereafter, three of the most recognized and researched technologies nowadays will be briefly presented, which are thus worthy of attention.

Industrial Internet Reference Architecture (*IIRA*) is a standards-oriented industrial internet system open architecture of the Industrial Internet Consortium (IIC), which aims to expand industry interoperability and guide the development of technical standards. IIRA is a highly abstract general-purpose architecture whose aim is to support a wide range of industrial applications based on use-cases defined by the IIC. The Architectural Framework (IIAF) is based on the ISO/IEC/IEEE 42010 (2011) standard specifications and codifies the conventions and common practices that are the core of the IIRA Internet Information Service (IIS). The architectural concepts include concerns (subjects of interest in the system), stakeholders (entities interested in the system), and viewpoints (conventions to construct a description and analysis of specific system concerns).

The core of Reference Architecture Model Industrie 4.0 (*RAMI 4.0*) is the concept of CPS (which is somehow similar to what IIS is in IIRA), where autonomy is localized and the component systems can make their own decisions. Its reference architecture model is more manufacturing-oriented and represents the integration of multiple stakeholder visions on how to realize I4.0-based on existing communication standards and functional descriptions. There are six vertical layers defining the nature of IT components in I4.0: business applications, functional aspects, information processing, comm unication, and integration capabilities, and the ability of assets to implement I4.0 features. The central concept of RAMI 4.0 concerns the I4.0 compliance of components through the deployment of Administration Shell (AS), which essentially provides a virtual representation of the entire life-cycle of an object or asset.

The Open Platform Communications Unified Architecture (*OPC UA*) is a product of the OPC Foundation ([51]), whose aim is to provide a service-oriented, platform-independent standard whose specification is organized into several documents, ranging from concepts and security model to services, data access, alarms, and so forth. OPC UA defines the relations and semantics of services that servers can provide, mapped onto different communication and transport protocols. Interoperability is enabled by mapping concepts between property sets from different domains and/or frameworks. There is an interesting research work presented in [16], which shows a service model similarity evidence and an interoperability affinity between OPC UA and RAMI 4.0 Administration Shell concepts.

2.5. Requirements of the Realized System

The MESS will be utilized to execute production activities in a manufacturing facility, as well as to enable service-oriented orchestration and production improvement with other external technologies. The aim of the system is not only to support and improve communication among the CPSs inside the facility, but also to improve the communication capability between the production and the other activities in the enterprise (such as process planning, process simulation, resource planning, etc.), to provide updated communication and information analysis to the management, and offer a clear and simple interface to the end-users, in order to monitor and operate the system. Briefly, it is expected of the implemented MESS to provide an interoperable, reconfigurable, and reliable information system to meet these requirements.

Several fundamental design considerations have been taken into account in specifying the basic functionality of the MESS, as listed hereafter:

(i) enabling both high- and low-level intelligent solutions in the production control hierarchy;
(ii) facilitating modern, standardized and easy-to-adopt integration on both end-points;
(iii) leveraging built-in asynchronous, message-based dispatching systems
(iv) propagating both historical and real-time data collection;
(v) obtaining a robust, process-focused, event-based and parallel-running control logic for the realization of a time and product "irrelevant" result;
(vi) utilizing service-oriented architectural approaches and IT solutions;

(vii) guaranteeing openness to predefined exception handling, as well as to unpredictable execution changing.

A careful selection of the required functionalities has been done for the achievement of an operation-focused and functionally minimalist, but useful MES. The manufacturing execution functions were decomposed according to their specific context and use of interfaces: production process management, data acquisition, and system monitoring. It is expected by the system to initiate, control, and report on production activities as they occur. To support these expectations, a summarized list of major (i.e., considered mandatory) and *core functionalities* for the MESS are catalogued in the following Table 2 based on the previously listed functionality collections.

Table 2. List of major functional requirements of the MESS.

Name	Description	Rationale
Production resource capabilities list	Provide a list of all of the resources available, together with their specific capabilities.	To have an overall view of what can or cannot be done in the system in terms of production capabilities combination.
Resource change management	Provide a preliminary analysis and validation workflow for changes in production resources. It aims at guiding the users along a process of testing, configuration and acceptance of the new production resource and its capabilities.	To identify early problems in production environment configuration.
Production data acquisition and collection	Monitor all resources of the production system, and store the acquired data for later reporting and analysis.	To acquire and collect operational data.
Production tasks dispatching	Address each single production step to the correct CPS and evaluate the overall answer.	To guarantee the link between control and physical execution.
Production process control	Be the system responsible to initiate and execute the list of tasks contained in the production process in the correct order.	To guarantee the correct execution of tasks sequence on the basis of their precedence.
Production process tracking	Monitor the progress of production and provide up-to-the-minute report on the production status.	To provide process supervision and tracking adherent to the real advancement of production process.
Production process planning	Provide an adaptive digital twin of the production process, which makes timely decisions to adjust the schedule and the process plan when unexpected situations occur.	To have a system component capable of managing requests to adapt process execution on the basis of inter-occurred environmental changes.

3. Methodology

The research work presented in this paper embodies a re-elaboration of the Networked Embedded Systems (NES) definition of a CPS [52,53] and its information and service model, with the intention to leverage it for the development and integration of I4.0-compliant CPSs in a distributed MES environment. This was also carried out following the RAMI 4.0 and OPC UA interoperability guidelines. To this end, the necessary industrial equipment and tools have been identified, whereas the CPS common service concepts and its core functionality has been outlined, as detailed in this section.

3.1. Basic Concepts

Manufacturing is fundamentally the process to plan and deploy an optimum way of transforming materials into goods, by means of integrating a variety of resources (people, capital, processes, systems, and enterprises), in order to deliver end-products of value to the market. Production starts with a product design. It always has a Bill of Materials (BoM), which can be translated into a *Process Plan*, which is a sequence of selected manufacturing or logistic related *Capabilities* to be performed by specific production *Resources* on the components of the BoM at the shop-floor level. A production task can be defined as the reservation of specific Resource based on its required Capability. Execution is the procedure to perform a Process Plan in the production system.

After defining the procedure of production, when a production order arrives, a planner can create a *Routing* by finding the feasible combination of available Resources with the necessary Capabilities to be performed, as well as the required materials and parts, together denominated as *Workpieces*. These one-by-one combinations of Resource, Capability, and Workpiece are called *Operations*, which are the basic units of a Routing. With a scheduler, a Routing's time course (called *Schedule*) can also be determined to finish before the order's deadline. The list of basic concepts leveraged in this work's methodology is reported in Table 3.

Table 3. List of manufacturing concepts utilized.

Name	Description
Resource	Any manufacturing or logistic machine/device or human operator in the production system that has the ability to perform production related activity.
Capability	Abstract description of the functionality that is provided by the Resources towards the system (e.g., gripping, drilling, identification of elements).
Task	A binding between specific Resources and a selected Capability to perform the required production related activity.
Process Plan	A sequence of selected Tasks necessary to perform specific complex production related activities, based on the technological precedence constraints (e.g., assembling, transporting, quality checking).
Workpiece	Every material, part, sub-assembly, assembly and product which is in the production system. (out of scope of this paper)
Operation	A binding between a specific Task and the Workpiece, which is the object of the specified Task. (out of scope of this paper)
Routing/Schedule	The sequence of selected Operations necessary to produce one or more products in the manufacturing system. The Schedule is the time-based dimension of the planned Routing. (out of scope of this paper)

3.2. *Cps Service Model and Architecture of the Mess*

At the heart of the developed system, there is the concept of CPS, which is in charge to perform production activities. As mentioned, the CPS conceptualization utilized in this manuscript is based on NES definition and comprises the following four basic abilities: sensing, acting, computing, and networking. A CPS can be abstracted to almost anything: a robot, a human worker with network connected device, a camera, a PLC controlled manufacturing unit, a pool of tools, etc. To the MESS, a CPS is like a black box that makes the physical layer seamless, providing a set of production related capabilities and variables. The production capabilities of the CPS can be reached by the production process manager (from now on MESS Core) directly or through a digital counterpart—digital twin—with standardized interface. From the point of view of an external component to the MESS (i.e., user interface, scheduler, and so forth, out of scope in this paper), only these twinned CPSs are visible and discoverable, while physical devices are kept hidden for security and competency reasons. A schematic representation of the system is illustrated in Figure 2.

In order to realize the idea of generally embeddable I4.0-compliant CPSs providing manufacturing services, a "minimalistic" CPS service model has been conceptualized. A CPS is basically expected to embody a set of core service concepts whose selection is necessary to guarantee the following: the core *digital representation* of a CPS; the *service interface* to the MESS collaborative environment; and the *compliance* of the CPS with NES definition and I4.0 components in RAMI 4.0 [16].

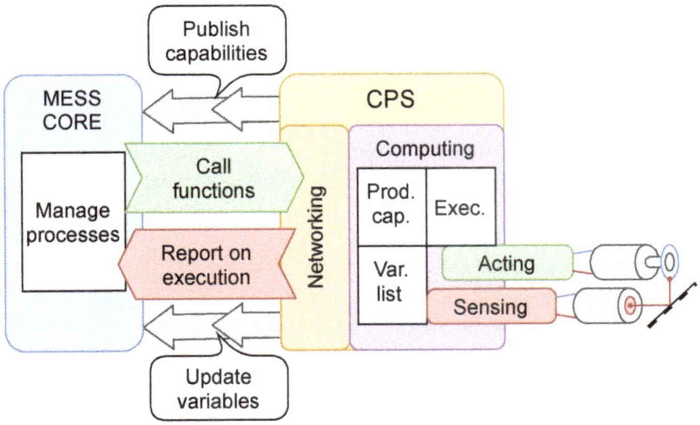

Figure 2. High-level concept of the MESS architecture.

All CPSs publish their capability in terms of (micro or macro) *services*, which can be invoked by means of parameterized *functions*. Invoking a function *call* triggers the instantiation of execution related tasks, which are necessary to track its execution advancement on the basis of even-triggered *reports*. CPSs also have operating parameters called *variables*, which can point to any exposed signal of the specific equipment and whose values can be utilized in the decision-making process of a routing. Functions are organized and linked together in a process plan via precedence edges, which represent the necessary conditions for the specific function to execute (details and an illustrative example in Section 5).

The interface of a CPS identifies the following common functionality:

(i) enables to connect, disconnect, and refresh requests from the system core;
(ii) provides information on CPS structure—i.e., device description, capabilities, and variables—and its statuses;
(iii) enables the execution of a function call;
(iv) reports on the call's status changes;
(v) provides the call's historical information;
(vi) reports changes on subscribed variable; and
(vii) provides error handling functions (cancel, end, kill process tasks).

Design assumptions:

(i) one CPS may contain one or more physical devices;
(ii) on the contrary, one device can only belong to exactly one CPS;
(iii) variables and functionality of the devices can be reached only through the CPS abstraction model; and
(iv) the CPS functions can be a logical combination of different physical devices' functions.

Any dedicated, low-level controller is allowed inside the CPS, but each CPS should have a unified interface toward the MESS. The definition of this minimal set of pillar CPS concepts is reported in Table 4.

3.3. Common OPC UA Model of a CPS

Figure 3 illustrates the concept map defined in the OPC UA information space, which is shared by a CPS when connected to the core engine of the MESS. The OPC UA address space is structured hierarchically to foster the interoperability of clients and servers, but only top levels are standardized for all servers (nodes in grey).

All other nodes in the address space follow this initial hierarchy but their specific information model is left to the system designers' freedom to interpret the environmental description. To resolve this initial model uncertainty, a set of common OPC UA nodes is proposed as the basis for every CPS service model. These nodes reflect the concepts introduced in the previous section and can be either statically or dynamically generated elements of the address space. For instance, objects organizing the overall OPC UA model structure are static (i.e., CPS, Functions, Calls, Variables, and Maintenance Functions), whereas OPC UA nodes related to the actual list of CPS functions, their call statuses, as well as the list of CPS variables are generated according to the specific configuration of the CPS and the execution logic of the process plan (dotted nodes). The core concept of a CPS variable should not be confused with the native OPC UA node of type *variable*, which is proper of the OPC UA specification. CPS and maintenance functions (yellow nodes) are OPC UA *methods*: the former expresses the service granularity of a CPS, while the latter are intended to provide the CPS administrator with a list of utilities regarding the overall process plan, the function calls, and the CPS configuration updates. The capabilities of a CPS follow the naming convention of the OPC UA technology and are specified and extended via a dedicated OPC UA namespace, which is structured hierarchically starting from the main CPS folder. The MESS can handle different reporting mechanisms for tracking the status of a CPS function call execution, but, by default, it requires that each function call generate a proper set of status reporting variables.

Table 4. List of CPS service concepts.

Name	Description
CPS Function	From the point of view of an external production environment, any consumable (micro or macro) service with a well-defined, perceivable utility in the overall process.
CPS Call	The sequence of actions and events after the invocation of a CPS Function on a specific CPS until it is carried out flawlessly.
CPS Report	Any feed-back or status update in the advancement of a CPS Call execution regarding production or environmental changes.
CPS Variable	Any observable physical signal or computed quantity produced by the CPS and published to the external world.

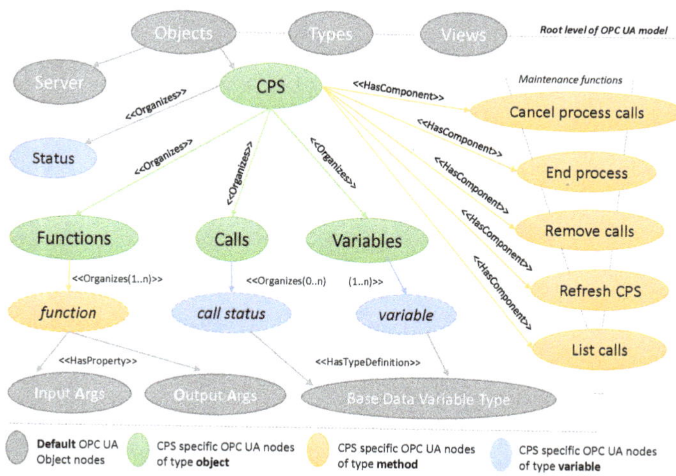

Figure 3. OPC UA information model of CPS core service elements.

4. System Implementation

The MESS is a set of integrated software and hardware components that provide functions for managing production activities, from job order launch to finished products. By the use of current and accurate data, the MESS initiates, guides, responds to, and reports on production activities as they occur (as expected by MESA). It can also interface with other production information systems, as well as support engineering and business activities (such as scheduling, simulation, process planning, and low-level task dispatching and process control on the physical layer of the CPS; all out of scope in this paper). The major architectural components of the MESS are depicted in Figure 4 and presented in the following sections.

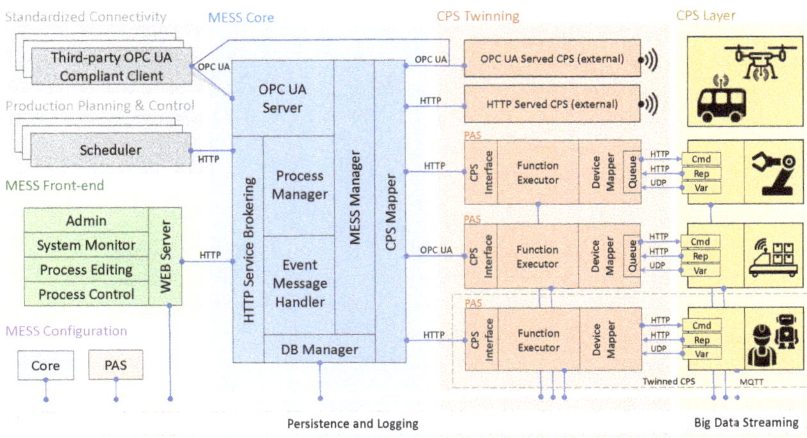

Figure 4. Overall view of the MESS architecture.

4.1. Mess Core

This is the heart of the system: a time-invariant, event-based sequence control of processes, based on the monitored statuses of resources (e.g., management of production processes). The MESS Core controls the synchronization with the CPSs and acknowledges the other system components about the occurring events. As illustrated in Figure 4, its major components can be listed as follows (from left to right):

- *HTTP Service Brokering Manager* for the front-end interface to the client components. It consists of an HTTP server to receive user requests, and an HTTP client to pass event messages to the connected clients;
- *OPC UA Server* for external OPC UA compliant client interface. It embodies and leverages an extended version of the Milo open-source project for complete OPC UA stack management;
- *Process Manager* is in charge for the execution of the process plans. It is one of the most crucial components of the MESS Core component. It manages the parallel execution of different process plans from different sources. It takes care of the proper locking and releasing mechanism of the resources. Nevertheless, it guarantees the precedence constraints between the operations (CPS functions) during the process plan execution, so that technologically irreplaceable operations cannot be interchanged. The sequence of CPS functions is to guarantee an acceptable and feasible production plan;
- *Event Message Handler* implements and manages the event-based logic of the MESS Core. Clients can sign up in order to monitor various kinds of events, and the manager is in charge to notify them both on the status changes of the MESS core and the CPS components, as well as on changes in the physical variables' value. It can also dispatch messages from the CPS;
- *Database Manager* is responsible for the persistent event logging. It records the events in the original form and manages the temporal labeling of the program, process, and operation life-cycles, as well;
- *MESS Manager* controls the communication between the other program layers;

- *CPS Mapper* is finally entitled to provide a unified mechanism to synchronize and consistently manage the connection and the information exchanged between the MESS Manager and the CPS Twinning layer (and its various CPS components).

The overall MESS also embodies a *Process Plan Editor*, which is part of the MESS Graphical User Interface (GUI). It provides a visual tool for the editing of process plan, to be executed by the MESS Process Manager later on. On the front-end side, the function inventory of every registered CPS is accessible and transparent with drag-and-drop feature to help users' interaction. Furthermore, the user interface provides a nearly real-time system overview during the process execution with the presentation of up-to-date production data. Beyond all of this, the admin users have a dedicated and secured interface for the configuration of the production and the MESS system.

4.2. CPS Twinning

As already mentioned, at the core of the designed system, there is the concept of CPS and the exploitation of its *Digital Twin* counterpart in (also virtually augmented) production scenarios. A tool can be any kind of manufacturing unit, an instrument, a device, a robot, a machine, a production line, a transport unit, or even a human worker. These tools are contained in (belong to) a CPS and the MESS Core takes care and manages the connections with all the CPSs. The MESS provides several ways to integrate CPSs and their tools into the system, according to the level of IT expertise and the communication control requested by their designers. Basically, it offers two interface solutions for this connection: by means of an externally served CPS (capable of one between OPC UA—the preferred solution—and HTTP/JSON equivalent service Application Programming Interface, in short API) or by the adoption of a lower-level wrapping Production Administration Shell (PAS) for CPS (part of this research work's contribution and described in detail in the following), respectively. The PAS represents the I4.0 compliant twinning counterpart of a generic production CPS, in which the latter is in control of the physical devices. The great advantage of this CPS Twinning solution is that, regardless of the specific method chosen by the designer for their CPS integration, once connected to the MESS Core, the overall system will seamlessly provide a standardized OPC UA-equivalent transformation of the CPS service and information model to the external world. It is worth recalling how the OPC UA is the technology sitting on top of the MESS (Core and CPS Twinning) standardization and semantic interoperability. It is the technological umbrella that is capable not only of transporting machine data but also semantically describing them in a machine-readable way. An example for the interoperability of a Twinned CPS is shown in Figure 5 through an external, commercial OPC UA client software.

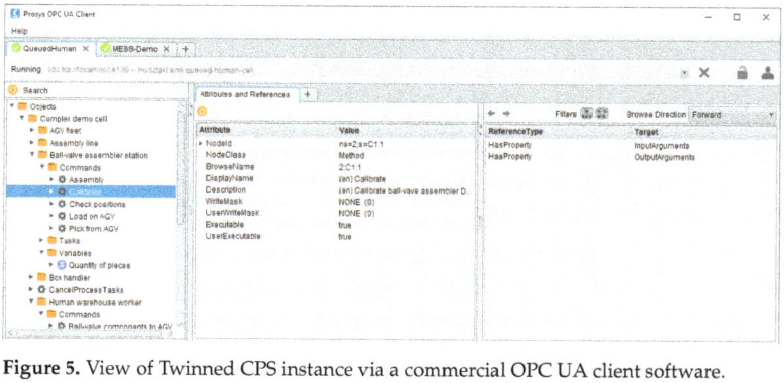

Figure 5. View of Twinned CPS instance via a commercial OPC UA client software.

The externally served CPS solution provides the opportunity to the CPS designers to create their own CPS controller with one of the MESS supported communication methods. The advantage of this solution is that the CPS developer is in charge of the complete implementation and communication, either via an OPC-UA server or an HTTP client and server pair.

The second solution is one of the peculiar outcomes of this research work and refers to the exploitation of the MESS PAS (details in the next section). Once its configuration is finalized, the PAS enables the construction of complex functionality thanks to the combination of the newly integrated tools' functions. The PAS internally models tools as devices, which can cover any kind of equipment, both individually or in a form of fleet (details in the demo use-case section).

4.3. Production Administration Shell

PAS takes inspiration from the design of the RAMI 4.0 AS (the similarity in representation can also be observed in Figure 6). Simply put, it represents a wrapping component that embodies, organizes, and exposes, in a standardized manner, the CPS functionality towards the overall production execution environment (refer to Figure 4 for architectural details). The goal of the PAS is to provide an easy-to-use tool which helps with transforming typically ad-hoc developed (legacy) manufacturing cells into interoperable, service-based I4.0 CPSs for distributed production environments. PAS embodies a dedicated OPC UA server (extension of the MILO Project, serving both the MESS core and third party OPC UA clients) and an HTTP server/client interface, which is capable of providing the same functionality and services defined for the MESS Core but specified/customized for the local intelligence and complexity of the single CPS.

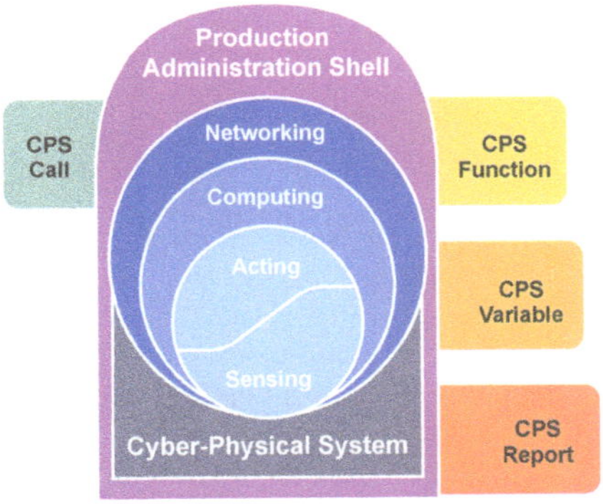

Figure 6. Production Administration Shell for CPS modeling and functionalities.

The PAS embodies a *Device Communication Layer* (DCL) to communicate with the physical devices, which is responsible for sending task execution requests, receiving reports on task execution statuses, sending special (maintenance) service requests to the devices (in order to query or cancel active tasks), as well as collecting the device variables' current values. Sending executable and maintenance tasks, as well as receiving inherent reports and variable updates from the CPS physical layer, can occur through different channels (HTTP or TCP), whereas intensive (Big Data) physical signal variables can be sent over UDP or MQTT, both to internal components or to external, cloud-based services for analytical purposes. As previously mentioned, the CPS Twinning layer of the MESS is able to generate a standardized OPC UA-equivalent transformation of the CPS information model. This is the core feature of the PAS too (details on the interfaces between the MESS Core, the CPS Twinning/PAS, and the low-level CPS communications can be found in Figure 7.

OPC UA was not chosen (although capable) for the intensive, low-level communications between the PAS and the CPS physical layer, due to the following reasons: first of all, the OPC UA is a complex information modeling technology to be deployed, and not just a communication and transport protocol; this can possibly result in communication stack overload and so response time issues at low-level, especially in resource-limited devices; in other words, OPC UA is not (primarily) intended for IoT bottom-up big data publishing (as also reported in [54]). To this end, the MQTT lightweight publish/subscribe messaging transport technology was selected for bottom-up data interoperability, even considering the possibility of pairing this technology with OPC UA in a publisher–subscriber model, eventually. Alternatively, the PAS has also built-in UDP data exchange interface.

Task execution requests can occur in two different modes, according to the specific device capabilities. By default, this layer handles task queues for the devices and dispatches tasks one-by-one. If a device is natively able to handle execution queues or to process more than one task at a time (what can be called *flooded* task execution), this setting can be omitted. The PAS also contains a built-in planner and a device controller, which provide the same realization of the OPC UA and HTTP interfaces proper of the MESS core but with the main goals to transform the high-level command tasks into low level, device-specific ones, and to control the execution of this new task graph.

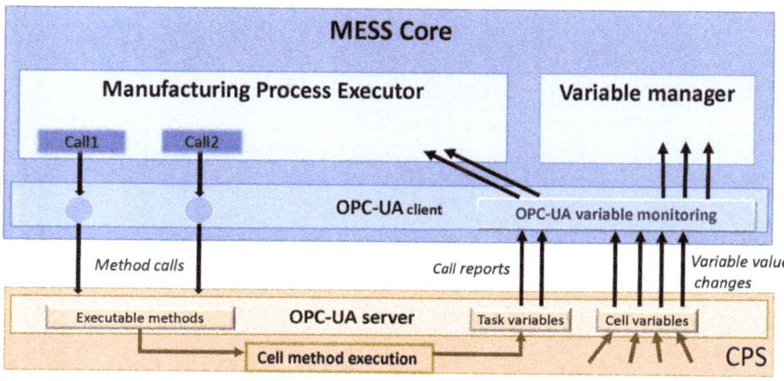

Figure 7. Details on the OPC UA interface between CPS and MESS Core.

The types of CPS utilized in this work comprise a variety of elements: single robot arms (1), production assembly line (2), Automated Guided Vehicle (AGV) fleet, with and without robot arms (3, 5), collaborative robots (4) and human operated components, such as the warehouse and a digital work assistance system (6). There is also a built-in *utility* CPS, whose *internal* capabilities are related to timing, monitoring, and changing conditions in the execution of process plans. A numbered correspondence of CPSs both for the planned and physically realized layout is reported in the following Figures 8 and 9.

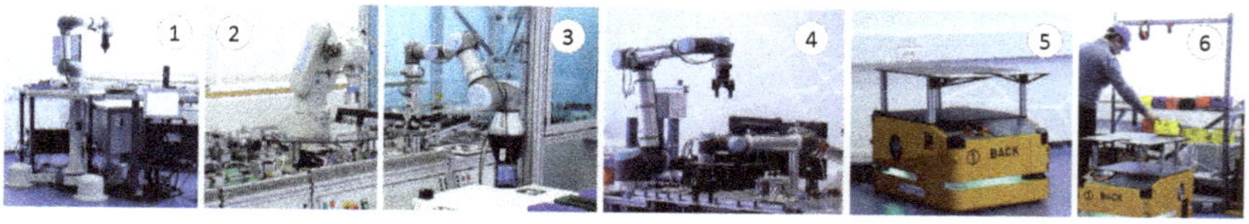

Figure 8. Physical components of the CPSs in the demonstrated use case.

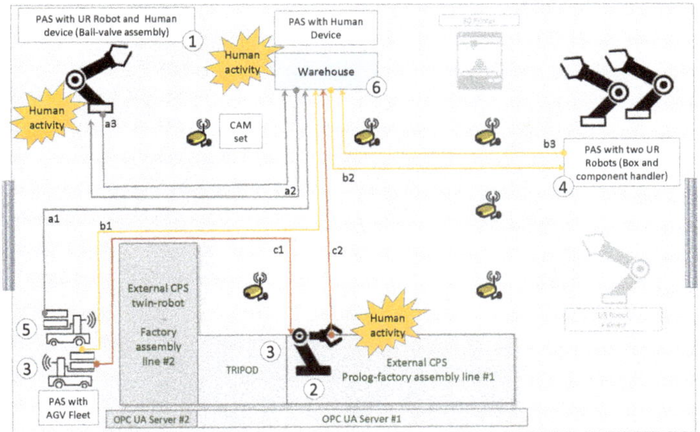

Figure 9. Realized MESS layout: involved CPSs, human activity, and implemented logistics.

5. Demonstration Use-Case

The MESS system was deployed in a pilot cyber-physical production and logistics system which was dedicated to research, education, training, and demonstrations in the academic, industrial, and public sectors [55]. In this *learning factory*, the goal was to verify and validate the MESS system and demonstrate its main services and facilities. The scenarios in the demonstration use-case were rather straightforward by design, in order to clearly exhibit the following features of the MESS: (1) interlinking of heterogeneous resources both (2) of production and of internal logistics; (3) inclusion of physical resources and their full-fledged digital twins; (4) integration of both legacy and state-of-the-art I4.0 ready, as well as (5) of fully automated and human-assisted system components; (6) and demonstration of their easy interoperability. Here, the emphasis is on the fact that some of the integrated elements—such as an assembly line [56], human-robot collaborative workcells [57], or a small fleet of AGVs [58]—were complete, complex CPSs themselves, accompanied by their specific, custom-tailored digital twins.

5.1. The Experimental Scenario

The use-case scenario presented here addresses the production (assembly and delivery) of ball–valves and thermometers in one of the MESS premises. The physical components of the pilot system participating in the use-case scenario are shown in Figure 8. The layout of the realized MESS, the sequence of HRC dynamics, and the involved logistics for the material handling (noted with $a1..a3$, $b1..b2$, $c1..c3$ routes) are depicted in Figure 9.

The main steps of the use-case scenario can be summarized as follows:

Step 1: Request of an AGV at the warehouse, where the human operator loads the pieces needed at the box handling cell;

Step 2: In parallel, another AGV is dispatched to the warehouse for the necessary ball–valve assembly pieces;

Step 3: AGVs deliver the materials to the docking positions indicated;

Step 4: The box-handling robot picks up the pieces, orders them, and releases the box by positioning it onto the AGV;

Step 5: At the ball–valve station, the human operator picks up the pieces, places them on the fixture, checks position and tools, and acknowledges back. Calibration and assembly operations then follow. When completed, the human operator loads the assembled ball–valve onto the AGV, which delivers it to the warehouse;

Step 6: The human operator at the warehouse unloads the manufactured pieces and stores them. The AGV is then released;

Step 7: At a certain time, another AGV is requested to the factory line #1, which is operating on the thermometers;
Step 8: When completed, the robot from factory line #1 will load the fixture containing thermometers onto the AGV, which will finally deliver them to the warehouse;
Step 9: As in point 6;
Step 10: AGVs are finally released and sent to the docking station for charging.

The list of targets and relative services implemented via OPC UA for each CPS in the demonstrated use-case scenario is reported in Table 5, whilst the graph in Figure 10 illustrates the concatenation of CPS services in the form of a schematic operation graph (routing), with their execution preconditions (precedences between nodes).

Table 5. List of published services for each CPS.

CPS	Integration Type	CPS Controllers	Production Services
Ball-valve assembly station	PAS with HTTP	Robot Controller Human-directed device	M1—Calibrate M2—Assembly ball-valve H1—Pick pieces from AGV, place them on fixture and check position H2—Check tools and robot add-ons correct positioning H3—Load assembled ball-valve onto AGV
Warehouse	PAS with OPC-UA	Human-directed device	H4—Check correct quantity of pieces in containers H5—Put ball-valve components on AGV H6—Pick and stock assembled ball-valve from AGV H7—Put components box on AGV H8—Pick and stock ordered box from AGV H9—Load fixture on AGV H10—Pick, stock pieces from AGV and return fixture to factory docking station
Assembly line	External OPC-UA cell	PLC controller Human operator	F1—Execute assembly order F2—Robot block (activates the secure handling zone of manufactured pieces by the human operator) F3—Robot unblock (de-activates the secure zone) H11—Control correct flow of assembled pieces
Box handler	PAS with HTTP	Box controller	C1—Pick up components from AGV (without robotic arm) C2—Manufacture pieces C3—Load final pieces onto AGV
AGV Fleet	PAS with HTTP	Fleet controller	AGV1—Go and wait to a predefined location AGV2—Transport components to a destination, through an intermediate cell AGV3—Continue to destination AGV4—Finish (indicates that transport is completed)

The previously mentioned *Process Plan Editor* provided by the MESS GUI is where, similarly to the schematic graph, the graph of a process plan can be created by selecting the available operations and connecting them with arrows to form precedence constraints. The process plan for this use-case scenario is shown in Figure 11. Additional information is available for every node derived from the common service model by clicking the info button, as shown in the blue box.

The created plan can be executed and its status can be tracked continuously on the *Process Manager* tab of the MESS GUI. The status of the execution can be visualized by the coloring of each individual task based on their own statuses. This is a quick and easy to understand approach to report on the overall status of the process. The coloring is the following: (i.) grey displays the task is waiting for start signal, (ii.) yellow means the task is under execution; meanwhile, (iii.) the green indicates that the task is finished and (iv.) the red color represents the error status. Except for the error status, the other three can be observed in Figure 12. The communication which reports on the shown status change is visible inside the box in the middle of the figure. The message contains the most necessary data to update the information model of the system: type of the message, process identifier, timestamp, and the type specific parameters. This lightweight communication between the CPS and the MESS Core allows a reliable operation. The detailed use-case scenario has been executed flawlessly in an automatic and trackable manner.

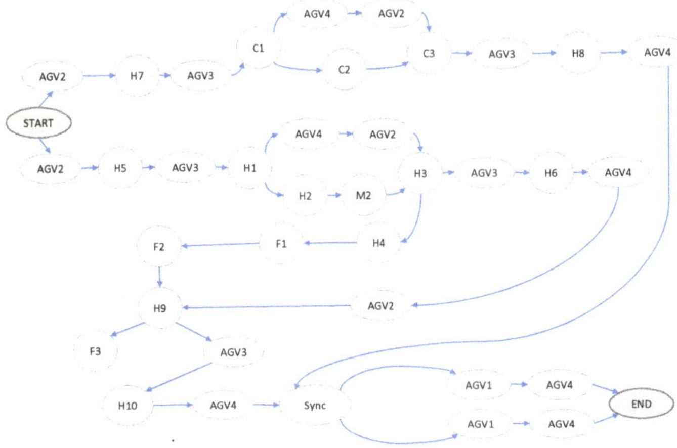

Figure 10. Schematic graph of the demonstrated use-case scenario.

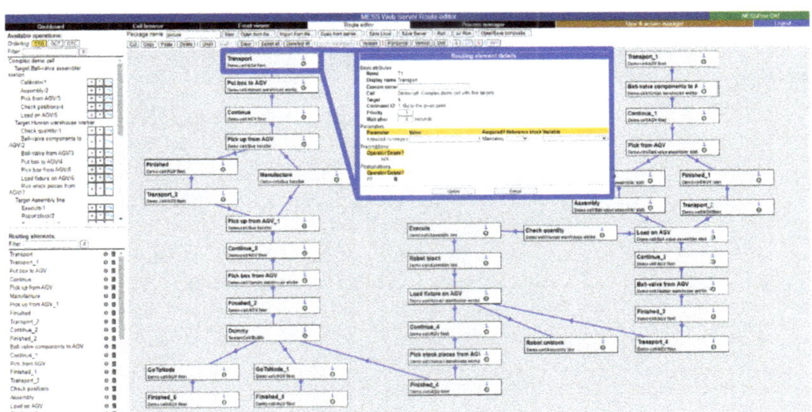

Figure 11. Process plan of the demonstrated use-case scenario.

5.2. *Discussion and Lessons Learned*

The above experimental setup and scenarios of the use-case proved to be appropriate to demonstrate all the above six features of the MESS. Hence, system components could be (de-)activated in a plug-and-play manner. Production and internal logistics could speak the same language, just like legacy robotic systems and brand new HRC workcells. Production and logistics processes could be run both in the virtual and physical worlds, even in parallel. This opens new opportunities both for right-first-time system design and configuration, as well as anticipating and avoiding failure situations. In settings where humans are closely involved in production, such a service is essential, whereas its lack could be detrimental [20,55].

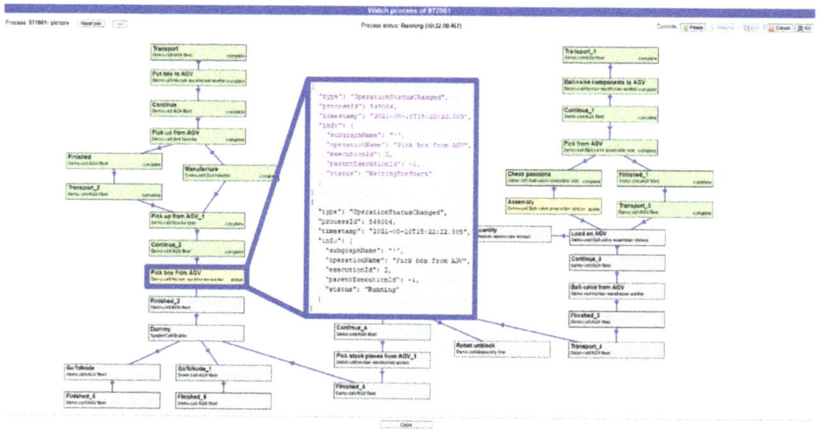

Figure 12. Process execution example with communication.

The implemented MESS and related (PAS-based) CPS resulted in a valid solution for an easier (re)configuration, as well as extensibility and management of the production and logistics processes, thanks to the conceptualized CPS common service model. This led to the conversion of isolated, legacy industrial equipment into I4.0 ready CPSs, which are capable of exposing their functionalities in terms of services. On the other hand, some CPSs evidenced specific issues. Factory line #1, for instance (no. 2 on Figure 8), could natively handle only a predefined number of *tasks*, and this number could not be changed dynamically during the operation. When parameterizing variables, moreover, they could only have read-only properties, due to system security and to avoid unwanted intervention. Nevertheless, by the protocol for writable variables, server-side and client-side writing were also allowed, which might cause unattended security risks. To overcome this, it was necessary to create cohesiveness between the field programmer level and the control programmer level. The establishing of the standardized synchronization of these programming levels requires further development and research work. It was also necessary, for safety reasons, to configure the network permissions in a way to restrict and route communication traffic only from accredited CPSs. After this initial setup, a variety of process plans were successfully carried out by the various CPSs, in a way that single production cells could be incrementally linked to much wider production processes.

6. Conclusions

Industry 4.0, or in general next generation of smart factory developments, will require an unprecedented level of interoperability and standardization, with the intent to facilitate the collaboration of connected cyber-physical entities. The latter, however, are all characterized by a high degree of flexibility, adaptability, and autonomy. In this paper, we suggested a generalized common service model and architecture of CPS-based manufacturing execution systems. The core model is minimalist as far as its underlying assumptions are concerned. Hence, it does not constrain the decision autonomy of collaborating cyber-physical entities, and "only" provides channels for transferring and synchronizing the information which ensue from their decisions. The proposed approach identified the elementary concepts, such as *functions, calls, variables,* and *reports* as the basis for modeling and providing I4.0 compliant, CPS-based services in a manufacturing environment. They have been developed via standardized technologies enabling semantic interoperability and openness (OPC UA, MQTT).

The universal CPS-based service integration mechanism has been validated in an experimental pilot production and logistics system which included a variety of heterogeneous and autonomous resources, such as manufacturing cells, AGVs, robots, and human–robot collaborative cells. These CPS components were connected and controlled via the plug and collaborate mechanism of our MESS system in a number of complex scenarios.

Next investigations look forward to the extension of the MESS in particular with including the semantics of the OPC UA common service model to support the adaptation and embodiment of new equipment, like 3D printers, palletizers, and taggers for internal positioning and logistics. The MESS system was prepared also with a distributed intelligent control in mind. Our future research will focus on realizing such a control scheme on the basis of the MESS architecture presented here.

Author Contributions: Conceptualization, R.B. and G.P.; methodology, R.B. and G.P.; software and system realization, B.H. and G.P.; validation, R.B., G.P. and B.H.; formal analysis, R.B. and J.V.; resources, R.B. and J.V.; data curation, G.P. and B.H.; writing—original draft preparation, R.B. and G.P.; writing—review and editing, B.H. and J.V.; visualization, R.B., G.P. and B.H.; supervision, J.V.; project administration, G.P. and J.V.; funding acquisition, J.V. All authors have read and agreed to the published version of the manuscript.

Funding: This research has been supported by the GINOP-2.3.2-15-2016-00002 grant on an "I4.0 research and innovation center of excellence" and by the ED_18-2-2018-0006 grant on a "Research on prime exploitation of the potential provided by the industrial digitalization".

Institutional Review Board Statement: Not applicable.

Informed Consent Statement: Not applicable.

Data Availability Statement: Not applicable.

Conflicts of Interest: The authors declare no conflict of interest. The funders had no role in the design of the study; in the collection, analyses, or interpretation of data; in the writing of the manuscript, or in the decision to publish the results.

Abbreviations

The following abbreviations are used in this manuscript:

AGV	Automated Guided Vehicle
AI	Artificial Intelligence
ANSI	American National Standards Institute
ANSI/ISA-95	International Standard for Enterprise-Control System Integration
API	Application Programming Interface
ARTI	Activity Resource Type Instance Reference Architecture
AS	Administration Shell
BoM	Bill of Materials
CIM	Computer Integrated Manufacturing
Cmd	Command
CPS	Cyber-Physical System
CPPS	Cyber-Physical Production System
DAI	Distributed Artificial Intelligence
DCL	Device Communication Layer
ERP	Enterprise Resource Planning
GUI	Graphical User Interface
HMS	Holonic Manufacturing System
HRC	Human–Robot Collaboration
HTTP	HyperText Transfer Protocol
I4.0	Industry 4.0
IIAF	Industrial Internet Architectural Framework
IIC	Industrial Internet Consortium
IIoT	Industrial Internet of Things
IIRA	Industrial Internet Reference Architecture

IIS	Industrial Internet System
IoT	Internet of Things
ISA	International Society of Automation
IT	Information Technology
JSON	JavaScript Object Notation
MAS	Multi-Agent System
MES	Manufacturing Execution System
MESA	Manufacturing Enterprise Solutions Association International
MESS	Manufacturing Execution System as a Service
MIES	Manufacturing Information and Execution System
MOM	Manufacturing Operations Management
MQTT	Message Queuing Telemetry Transport
OPC UA	Open Platform Communications Unified Architecture
NIST	National Institute of Standards and Technology
NES	Networked Embedded System
PAS	Production Administration Shell
PLM	Plant Lifecycle Management
PROSA	Product-Resource-Order-Staff (Reference) Architecture for HMS
RAMI 4.0	Reference Architecture Model for Industry 4.0
Rep	Report
SEMATECH	Semiconductor Manufacturing Technology Consortium
SIMA	Systems Integration of Manufacturing Applications Reference Architecture
TCP	Transmission Control Protocol
UDP	User Datagram Protocol
Var	Variable
VDI	Verein Deutscher Ingenieure (translation: Association of German Engineers)
ZVEI	Zentralverband Elektrotechnik- und Elektronikindustri eV (translation: Electrical and Electronic Manufacturers' Association)

References

1. Monostori, L.; Váncza, J. Towards living manufacturing systems. *Procedia CIRP* **2020**, *93*, 323–328. [CrossRef]
2. Kletti, J. (Ed.) *Manufacturing Execution System—MES*; Springer: Berlin/Heidelberg, Germany, 2007. [CrossRef]
3. Monostori, L. Cyber-physical Production Systems: Roots, Expectations and R&D Challenges. *Procedia CIRP* **2014**, *17*, 9–13. [CrossRef]
4. Monostori, L.; Kádár, B.; Bauernhansl, T.; Kondoh, S.; Kumara, S.; Reinhart, G.; Sauer, O.; Schuh, G.; Sihn, W.; Ueda, K. Cyber-physical systems in manufacturing. *CIRP Ann.* **2016**, *65*, 621–641. [CrossRef]
5. Ezell, S. Why Manufacturing Digitalization Matters and How Countries Are Supporting It? 2018. Available online: http://www2.itif.org/2018-manufacturing-digitalization.pdf (accessed on 30 March 2020).
6. Kilimis, P.; Zou, W.; Lehmann, M.; Berger, U. A Survey on Digitalization for SMEs in Brandenburg, Germany. *IFAC PapersOnLine* **2019**, *52*, 2140–2145. [CrossRef]
7. Ignat, V. Digitalization and the global technology trends. In Proceedings of the 32nd International Conference on Computer Applications in Industry and Engineering, Sibiu, Romania, 14–17 June 2017; Volume 227, p. 012062. [CrossRef]
8. Muhamad Ibrahim, N.; Hassan, M.F. A Comprehensive Comparative Study of MOM for Adaptive Interoperability Communications in Service Oriented Architecture. *Int. J. Trend Sci. Res. Dev.* **2019**, *3*, 23–30. [CrossRef]
9. West, D.M. What Happens If Robots Take the Jobs? The Impact of Emerging Technologies on Employment and Public Policy. 2015. Available online: https://www.brookings.edu/wp-content/uploads/2016/06/robotwork.pdf (accessed on 20 April 2020).
10. i-SCOOP. Manufacturing Execution Systems (MES)—Evolutions and Software Solutions. Available online: https://www.i-scoop.eu/industry-4-0/manufacturing-execution-systems-mes-evolutions-software-solutions/ (accessed on 24 March 2020).
11. Stark, R.; Damerau, T. Digital Twin. In *CIRP Encyclopedia of Production Engineering*; Springer: Berlin/Heidelberg, Germany, 2019; pp. 1–8. [CrossRef]
12. Report, M.S. The Multi-Directional Evolution of MES Software. 2018. Available online: https://www.sme.org/technologies/articles/2018/june/the-multi-directional-evolution-of-mes-software/ (accessed on 20 August 2020).
13. MESA International. MESA Model V2.1. 2019. Available online: http://www.mesa.org/en/modelstrategicinitiatives/MESAModel.asp (accessed on 30 June 2020).

14. VDI; VDE; ZVEI. GMA Status Report: Reference Architecture Model Industrie 4.0 (RAMI 4.0). 2015. Available online: https://www.zvei.org/fileadmin/user_upload/Presse_und_Medien/Publikationen/2016/januar/GMA_Status_Report__Reference_Archtitecture_Model_Industrie_4.0__RAMI_4.0_/GMA-Status-Report-RAMI-40-July-2015.pdf (accessed on 30 June 2020).
15. Industrial Internet Consortium. Industrial Internet Reference Architecture. 2015. Available online: https://www.iiconsortium.org/IIRA.htm (accessed on 23 April 2018).
16. Pedone, G.; Mezgár, I. Model similarity evidence and interoperability affinity in cloud-ready Industry 4.0 technologies. *Comput. Ind.* **2018**, *100*, 278–286. [CrossRef]
17. Monostori, L.; Váncza, J.; Kumara, S.R. Agent-based systems for manufacturing. *CIRP Ann.* **2006**, *55*, 697–720. [CrossRef]
18. Enterprise Information Systems. INDUSTRY 4.0 STANDARDS. 2017. Available online: http://i40.semantic-interoperability.org/ (accessed on 30 June 2020).
19. Almada-Lobo, F. The Industry 4.0 revolution and the future of Manufacturing Execution Systems (MES). *J. Innov. Manag.* **2016**, *3*, 16–21. [CrossRef]
20. Kemény, Z.; Váncza, J.; Wang, L.; Wang, X.V. Human–Robot Collaboration in Manufacturing: A Multi-agent View. In *Advanced Human-Robot Collaboration in Manufacturing*; Wang, L., Wang, X.V., Váncza, J., Kemény, Z., Eds.; Springer International Publishing: Cham, Switzerland, 2021; pp. 3–41.
21. Mantravadi, S.; Møller, C. An Overview of Next,-generation Manufacturing Execution Systems: How important is MES for Industry 4.0? *Procedia Manuf.* **2019**, *30*, 588–595. [CrossRef]
22. Alcácer, V.; Cruz-Machado, V. Scanning the Industry 4.0: A Literature Review on Technologies for Manufacturing Systems. *Eng. Sci. Technol. Int. J.* **2019**, *22*, 899–919. [CrossRef]
23. Lu, Y.; Xu, X. Resource virtualization: A core technology for developing cyber-physical production systems. *J. Manuf. Syst.* **2018**, *47*, 128–140. [CrossRef]
24. Cardin, O. Classification of cyber-physical production systems applications: Proposition of an analysis framework. *Comput. Ind.* **2019**, *104*, 11–21. [CrossRef]
25. Trabesinger, S.; Pichler, R.; Schall, D.; Gfrerer, R. Connectivity as a prior challenge in establishing CPPS on basis of heterogeneous IT-software environments. *Procedia Manuf.* **2019**, *31*, 370–376. [CrossRef]
26. Faller, C.; Höftmann, M. Service-oriented communication model for cyber-physical-production-systems. *Procedia CIRP* **2018**, *67*, 156–161. [CrossRef]
27. Biesinger, F.; Meike, D.; Kraß, B.; Weyrich, M. A digital twin for production planning based on cyber-physical systems: A Case Study for a Cyber-Physical System-Based Creation of a Digital Twin. *Procedia CIRP* **2019**, *79*, 355–360. [CrossRef]
28. Wunck, C. Towards a Microservice Architecture for the Manufacturing Operations Layer. In Proceedings of the 32nd International Conference on Computer Applications in Industry and Engineering, San Diego, CA, USA, 30 September–2 October 2019; Volume 63, pp. 241–250. [CrossRef]
29. García, M.V.; Irisarri, E.; Pérez, F.; Estévez, E.; Marcos, M. An Open CPPS Automation Architecture based on IEC-61499 over OPC-UA for flexible manufacturing in Oil&Gas Industry. *IFAC-PapersOnLine* **2017**, *50*, 1231–1238. [CrossRef]
30. Yang, C.; Lan, S.; Wang, L.; Shen, W.; Huang, G. Big data driven edge-cloud collaboration architecture for cloud manufacturing: A software defined perspective. *IEEE Access* **2020**, *8*, 45938–45950. [CrossRef]
31. Jo, G.; Jang, S.H.; Jeong, J. Design and Implementation of CPPS and Edge Computing Architecture based on OPC UA Server. *Procedia Comput. Sci.* **2019**, *155*, 97–104. [CrossRef]
32. Kim, J.; Jo, G.; Jeong, J. A Novel CPPS Architecture Integrated with Centralized OPC UA server for 5G-based Smart Manufacturing. *Procedia Comput. Sci.* **2019**, *155*, 113–120. [CrossRef]
33. Regal, T.; Pereira, C.E. Towards a conceptual model of structural and behavioral elements in cyber-physical production systems. *IFAC-PapersOnLine* **2019**, *52*, 863–868. [CrossRef]
34. Liu, C.; Jiang, P.; Jiang, W. Web-based digital twin modeling and remote control of cyber-physical production systems. *Robot. Comput. Integr. Manuf.* **2020**, *64*, 101956. [CrossRef]
35. Engelsberger, M.; Greiner, T. Dynamic reconfiguration of service-oriented resources in cyber—Physical production systems by a process-independent approach with multiple criteria and multiple resource management operations. *Future Gener. Comput. Syst.* **2018**, *88*, 424–441. [CrossRef]
36. Russell, S.; Wefald, E. *Do the Right Thing—Studies in Limited Rationality*; MIT Press: Cambridge, MA, USA, 1991.
37. Shih, W.; Srihari, K. Distributed Artificial Intelligence in manufacturing systems control. *Comput. Ind. Eng.* **1995**, *29*, 199–203. [CrossRef]
38. Arinez, J.F.; Chang, Q.; Gao, R.X.; Xu, C.; Zhang, J. Artificial Intelligence in Advanced Manufacturing: Current Status and Future Outlook. *J. Manuf. Sci. Eng.* **2020**, *142*, 110804. [CrossRef]
39. Leitao, P. Multi-agent Systems in Industry: Current Trends & Future Challenges. In *Beyond Artificial Intelligence. Topics in Intelligent Engineering and Informatics*; Kelemen, J., Romportl, J., Z.E., Eds.; Springer: Berlin/Heidelberg, Germany, 2013; Volume 4. [CrossRef]
40. Van Brussel, H.; Wyns, J.; Valckenaers, P.; Bongaerts, L.; Peeters, P. Reference architecture for holonic manufacturing systems: PROSA. *Comput. Ind.* **1998**, *37*, 255–274. [CrossRef]

41. Foit, K.; Banas, W.; Gwiazda, A.; Hryniewicz, P. The comparison of the use of holonic and agent-based methods in modeling of manufacturing systems. *IOP Conf. Ser. Mater. Sci. Eng.* **2017**, *227*, 012046. [CrossRef]
42. Saint Germain, B.; Valckenaers, P.; Van Brussel, H.; Van Belle, J. Networked manufacturing control: An industrial case. *Cirp J. Manuf. Sci. Technol.* **2011**, *4*, 324–326. [CrossRef]
43. Ádám S.; Kádár, B. Platform and direct exchange-based mechanisms for resource sharing in distributed manufacturing: A comparison. *CIRP Ann.* **2021**, *70*, 407–410. [CrossRef]
44. Ádám, S.; Pedone, G.; Egri, P.; Ádám, S.; Nick, G. A mutualistic framework for sustainable capacity sharing in manufacturing. *Procedia CIRP* **2020**, *93*, 938–943. [CrossRef]
45. Beregi, R.; Pedone, G.; Preuveneers, D. Towards trustworthy Cyber-physical Production Systems: A dynamic agent accountability approach. *J. Ambient. Intell. Smart Environ.* **2021**, *13*, 157–180. [CrossRef]
46. MESA International—White Paper #6. In *MES Explained: A High Level Vision*; Technical Report; MESA: Pittsburgh, PA, USA, 1997.
47. Computer Security Resource Center, Information Technology Laboratory, NIST. Glossary. 2021. Available online: https://csrc.nist.gov/glossary/term/Manufacturing_Execution_System (accessed on 31 May 2021).
48. Barkmeyer, E.; Christopher, N.; Feng, S.; Fowler, J.; Frechette, S.; Jones, A.; Jurrens, K.; McLean, C.; Pratt, M.; Scott, H.; et al. *SIMA Reference Architecture Part I: Activity Models*; NIST: Gaithersburg, MD, USA, 1987. [CrossRef]
49. VDI 5600 Blatt 1. *Fertigungsmanagementsysteme (Manufacturing Execution Systems—MES)*; Technical Report; VDI: Düsseldorf, Germany, 2016.
50. Hawker, J.S. CIM Framework architecture and application models. In *Information Infrastructure Systems for Manufacturing II: IFIP TC5 WG5.3/5.7, Proceedings of the Third International Working Conference on the Design of Information Infrastructure Systems for Manufacturing (DIISM'98), Fort Worth, TX, USA, 18–20 May 1998*; Springer US: Boston, MA, USA, 1999; pp. 201–214. [CrossRef]
51. OPC Foundation. OPC Unified Architecture. 2015. Available online: https://opcfoundation.org/developer-tools/specifications-unified-architecture (accessed on 19 May 2021).
52. Edward A.L.; Sanjit, A.S. *Introduction to Embedded Systems, A Cyber-Physical Systems Approach*, 2nd ed.; MIT Press: Cambridge, MA, USA, 2017.
53. Mosterman, P.J.; Zander, J. Cyber-physical systems challenges: A needs analysis for collaborating embedded software systems. *Softw. Syst. Model.* **2016**, *15*, 5–16. [CrossRef]
54. OPC Foundation. OPC Unified Architecture. Interoperability for Industrie 4.0 and the Internet of Things. 2019. Available online: https://www.festo.com/rep/en-gb_gb/assets/pdf/GB-OPC-UA-Interoperability-For-Industrie4-and-IoT-EN-v5-June-2017.pdf (accessed on 30 June 2020).
55. Kemény, Z.; Beregi, R.; Tipary, B.; Abai, K.; Nacsa, J. Recent advances in learning content and infrastructure development for layout and process planning courses at the SZTAKI learning factories. *Procedia Manuf.* **2020**, *45*, 319–324. [CrossRef]
56. Szántó, N.; Pedone, G.; Monek, G.; Háy, B.; Jósvai, J. Transformation of traditional assembly lines into interoperable CPPS for MES: An OPC UA enabled scenario. *Procedia Manuf.* **2021**, *54*, 118–123. [CrossRef]
57. Kemény, Z.; Beregi, R.; Nacsa, J.; Kardos, C.; Horváth, D. Human–robot collaboration in the MTA SZTAKI learning factory facility at Győr. *Procedia Manuf.* **2018**, *23*, 105–110. [CrossRef]
58. Kis, K.B.; Csempesz, J.; Csáji, B.C. A simultaneous localization and mapping algorithm for sensors with low sampling rate and its application to autonomous mobile robots. *Procedia Manuf.* **2021**, *54*, 154–159. [CrossRef]

applied sciences

MDPI

Article

Decentralized Multi-Agent Control of a Manipulator in Continuous Task Learning

Asad Ali Shahid [1], Jorge Said Vidal Sesin [2], Damjan Pecioski [2], Francesco Braghin [2], Dario Piga [1] and Loris Roveda [1,*]

[1] Istituto Dalle Molle di Studi Sull'Intelligenza Artificiale (IDSIA), Scuola Universitaria Professionale della Svizzera Italiana (SUPSI), Università della Svizzera Italiana (USI), CH-6962 Lugano-Viganello, Switzerland; asadali.shahid@idsia.ch (A.A.S.); dario.piga@supsi.ch (D.P.)

[2] Department of Mechanical Engineering, Politecnico di Milano, 20156 Milano, Italy; jorgesaid.vidal@mail.polimi.it (J.S.V.S.); damjan.pecioski@mail.polimi.it (D.P.); francesco.braghin@polimi.it (F.B.)

* Correspondence: loris.roveda@idsia.ch

Citation: Shahid, A.A.; Sesin, J.S.V.; Pecioski, D.; Braghin, F.; Piga, D.; Roveda, L. Decentralized Multi-Agent Control of a Manipulator in Continuous Task Learning. *Appl. Sci.* **2021**, *11*, 10227. https://doi.org/10.3390/app112110227

Academic Editor: Dario Richiedei

Received: 2 October 2021
Accepted: 27 October 2021
Published: 1 Novermber 2021

Abstract: Many real-world tasks require multiple agents to work together. When talking about multiple agents in robotics, it is usually referenced to multiple manipulators in collaboration to solve a given task, where each one is controlled by a single agent. However, due to the increasing development of modular and re-configurable robots, it is also important to investigate the possibility of implementing multi-agent controllers that learn how to manage the manipulator's degrees of freedom (DoF) in separated clusters for the execution of a given application (e.g., being able to face faults or, partially, new kinematics configurations). Within this context, this paper focuses on the decentralization of the robot control action learning and (re)execution considering a generic multi-DoF manipulator. Indeed, the proposed framework employs a multi-agent paradigm and investigates how such a framework impacts the control action learning process. Multiple variations of the multi-agent framework have been proposed and tested in this research, comparing the achieved performance w.r.t. a centralized (i.e., single-agent) control action learning framework, previously proposed by some of the authors. As a case study, a manipulation task (i.e., grasping and lifting) of an unknown object (to the robot controller) has been considered for validation, employing a Franka EMIKA panda robot. The MuJoCo environment has been employed to implement and test the proposed multi-agent framework. The achieved results show that the proposed decentralized approach is capable of accelerating the learning process at the beginning with respect to the single-agent framework while also reducing the computational effort. In fact, when decentralizing the controller, it is shown that the number of variables involved in the action space can be efficiently separated into several groups and several agents. This simplifies the original complex problem into multiple ones, efficiently improving the task learning process.

Keywords: reinforcement learning; decentralized control; multi-agent; continuous control; robotic grasping; policy optimization

1. Introduction

1.1. Context

Artificial intelligence and machine learning are quickly becoming core tools in the robotics domain, allowing incorporation of intelligence in robots, making them able to analyze, learn, and improve their behavior with minimum human intervention [1,2]. In particular, reinforcement learning (RL) [3] offers great potential to benefit from the experience to learn control tasks autonomously, as it is successfully shown in complex environments and scenarios [4,5]. However, reinforcement learning typically requires collecting a large number of samples to accumulate such an experience (i.e., training data), which still limits the application of RL in real physical systems such as robots, and pushes the current computer technology to the limit. Moreover, the control problems in robotics are best represented with high-dimensional matrices, continuous state, and action spaces, which actually make the application of this type of technique expensive even for highly efficient processors [6]. Therefore, it is important to consider new ways to simplify the learning problem, to reduce the computational efforts required by the RL. Recent works have considered ways to speed up the training process by learning contextual affordances [7], where the robot learns to perceive the effect of performing an action on an object using the current state. This is particularly important for learning vision-based manipulation skills since it greatly reduces the training time [8]. Some other works have instead focused on the explainability of RL agents, allowing a learning agent to explain the decision of selecting an action over other possible actions [9]. Exploration is another main challenge in RL agents that refers to the problem of choosing a policy that allows a learning agent to discover high-reward regions of the state space. In [10], the performance of different exploration strategies have been compared for on-policy and off-policy algorithms showing that the value-difference-based exploration strategy combined with softmax action selection improve the learning efficiency. Providing interactive feedback about the agent's actions is another approach to deal with the exploration issue [11].

In this paper, a decentralized learning approach is proposed for the robot to learn to perform a target task in a simulation environment through the use of a multi-agent framework (being implemented as On or Off policy RL algorithms), and transfer such knowledge to the real robot for real-task execution. Results and achieved performance are then analyzed in terms of how this approach can improve the computational effort and learning curve speed, while processing the learning algorithms, by comparing the proposed multi-agent framework with a centralized learning schema (used as a baseline).

1.2. State of the Art and Related Works

Within the state-of-the-art methods related to robotic task learning, the two main learning approaches that can be identified are (i) reinforcement learning and (ii) learning from demonstration (LfD) or imitation learning. The main difference between these approaches is that RL discovers optimal behavior through trials, whereas LfD aims to learn the task based on the presented trajectories, such as in [12]. A combination of these approaches can be used as presented in [5], where an algorithm learned to play the game of Go and won against an international champion by first learning from playing games and later improving through RL. In order to implement autonomous task-learning approaches, RL is commonly adopted. Recent works have emerged in the use of deep RL techniques in robotic grasping as in [13], and more notably in [14]. In [15], deep RL is used in the execution of a robotic task in a continuous domain. This is of great importance, since most real-life robots are simulated in continuous space. In [16], major advancements in large convolutional neural networks used for deep RL are highlighted. These networks have been used to learn hand-eye coordination for robotic grasping from monocular images and can predict the probability of successful grasping using only camera images.

With the increasing complexity of systems and tasks, a new efficient learning method, decentralized RL, has been proposed [17]. Additional work in this field is presented in [18], where decentralized reinforcement learning is implemented in learning robot behavior. Within the field of multi-agent systems, ref. [19] provides a survey for cooperative multi-agent learning literature where the work is divided into two categories. These categories are applying a single learner to discover joint solutions to multi-agent problems (team learning), or using multiple simultaneous learners, often one per agent (concurrent learning). The complexity of continuous space made the first algorithms used in RL obsolete, presenting the need for creating new approaches. These new approaches include networks such as Hindsight Experience Replay, which uses deep Q-networks (DQN) and deterministic gradients (DDPG), which can be found in [20]. Another approach is a distributed Q-learning algorithm which was introduced in [21]. It is noted that by using this algorithm, agents might neglect a penalty due to non-coordination in their update.

Ref. [22] points out the limitations of distributed Q-learning when dealing with stochastic environments, presenting a modified exploration strategy. Additionally, the Hysteretic Q-Learning algorithm [23] can be used for decentralized reinforcement learning. This algorithm is an extension of Q-learning and does not require any additional communication between the robots.

When studying multi-agent systems, the learning environment might be non-Markovian from an agent's perspective. This property is explored in [24], where the multi-agent systems are investigated from an evolutionary dynamical perspective. In [25], some known intuitions about concurrent learning agents are formalized, providing formal conditions that make the environment non-Markovian from an independent (non-communicative) learner's perspective. Cooperative control can be used to improve both the quality and speed of learning [26]. Additional parameterized control policies, which are used to reduce the computational load of high dimensional continuous state-action spaces, are presented in [27]. Ref. [28] combines deep learning with multi-agent systems, developing a deep multi-agent RL approach for decentralized continuous cooperative control. In [29], the distributed control of mobile robots is presented for environmental sensing applications. Closed-loop locomotion policies are learned for articulated mobile robots (snake and hexapod) using the multi-agent extension of asynchronous advantage actor-critic (A3C) algorithm [30]. The A3C algorithm acts as a global meta-agent storing a shared policy that multiple worker agents implement and improve in a distributed manner. Coordinating manipulation skills using two robots is also presented in [31], where a multi-agent approach using skill behavior diversification is implemented. The learning of skills is done separately and put together in order to achieve a complex task execution. In [32], a decentralized multi-arm motion planner is implemented using up to 10 robotic arms. The decentralized policy is trained to control one robot arm in the multi-arm system to reach its target end-effector position, given the observations of its workspace state.

1.3. Paper Contribution

The main objective of this work is to decentralize a centralized controlled system of a multi-DoF robotic arm. By decentralizing and splitting the actions into multiple separate agents, the proposed approach gives modularity to the system. Specifically, in the presented approach, two agents are used in the multi-agent system. The number of used agents, as well as the number of motors that a single agent controls, can be varied, providing flexibility and generalizible capabilities to the framework. Multiple variations of the recent RL algorithms (i.e., Proximal Policy Optimization—PPO—and Soft Actor-Critic—SAC) are used for the training. In the first case, the agents are coordinated by an overseer (a meta-agent), which gives importance to the actions of one of the two agents. In the other case, the agents are not coordinated, and instead, both the agents follow their own policy, either the PPO policy, the SAC policy, or the combination of both. In the uncoordinated case, the agents are not given any information about the output of the other agent, thus, not knowing that their actions affect the other agent and the overall system. In the final case, a single agent in a decentralized system is used. The manipulator's action are divided into two parts, and a single agent following one policy is used to learn the task (as shown in [33]).

In order to evaluate the proposed multi-agent framework, a 7-DoF Franka EMIKA panda robot is employed to perform a grasping task. The robot is required to position the end-effector above the object to be grasped, grasp the object, and finally lift it off of the surface. The overall performance of the grasping task is compared with the centralized system proposed in [33]. In more detail, the decentralization of the system is done by splitting the motors of the system into two equal parts, and giving the control of one part of the system to one agent and the rest to the other agent. Specifically, one agent controls the first four motors of the manipulator, while the second agent controls the last three motors of the manipulator and the gripper. The information of the system is shared between the agents, each one following a specific policy, with four output control actions that affect the next state. Training with multiple agents demonstrates the modularity of the process. The learned behavior can then be transferred to the real robot, to execute the target task and prove the scalability of the algorithm.

All of the achieved results show an acceleration in the learning process at the beginning of the episodes, even though the multi-agent approach leads to a stagnation of the learning curve. It is also shown that (with the correct setup of the reward function, the correct combination of On and Off policy algorithms, the correct number of agents, and the decentralization of the action space) a fewer amount of episodes are required to achieve the same reward of the baseline algorithm in [33] during the execution of the target manipulation task. This leads to an improvement of the overall training wall clock time, in combination with the efficiency gained for processing. The main scheme of the proposed multi-agent framework for the manipulator is shown in the Figure 1.

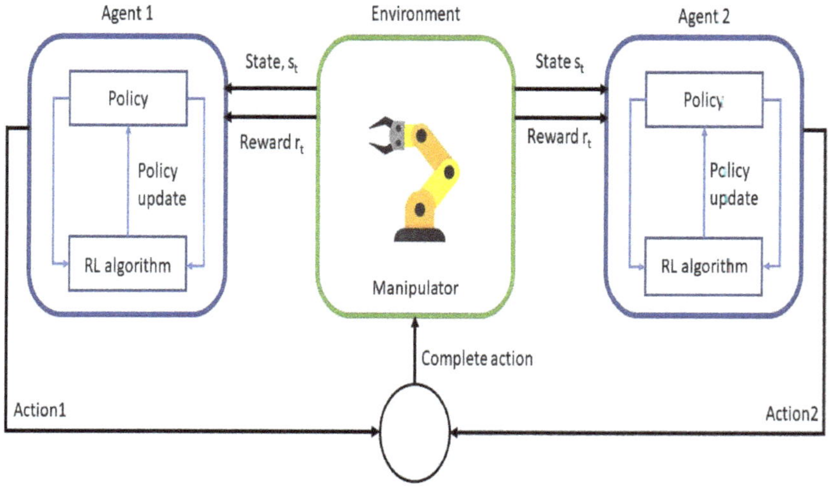

Figure 1. Main conceptual scheme of the proposed approach using a multi-agent and decentralized reinforcement learning approach.

1.4. Paper Layout

The paper is organized as follows. In Section 2, the addressed problem is described and principal challenges faced in the paper are explained, along with a schematic representation of the developed solution. Section 3 serves as an introduction to the methodology behind some of the key concepts of the proposed multi-agent framework decentralizing the robots control action learning. It provides details about the multi-agent and decentralized RL, together with a more detailed scheme of the key algorithms that have been employed. Section 4 presents the implementation of the proposed solution, providing details related to the learning environment, the training, and the defined reward function. In Section 5, a deep analysis of the experimental evaluation is provided, comparing the decentralized and multi-agent approach with the centralized algorithm in [33], which has been used as a baseline. The obtained results are discussed in Section 6. Finally, Section 7 states the conclusions, highlighting the limitations of the proposed approach, together with further possible future works to address them.

2. Problem Description

The primary objective of this paper is two-fold: (1) to provide an approach for decentralizing a single manipulator into multiple parts to show the possibility of using multiple agents in modular robots, and (2) to compare the results of the decentralized framework with the centralized framework proposed in [33]. To do that, the robot needs to be simulated in continuous space, as most of the real-world robotics problems are represented in this way. For such a purpose, the Mujoco multiphysics simulation engine has been employed as it is better detailed in the following sections. Multiple variations of model-free reinforcement learning algorithms are employed in the learning process. The on-policy algorithm, Proximal Policy Optimization (PPO), off-policy algorithm, and Soft Actor-Critic (SAC) are proposed to train the robot controller. Another algorithm that is proposed is the use of a meta-controller that coordinates the multiple agents in the system.

In order to implement, validate, and test the proposed multi-agent framework, a manipulation task (i.e., grasping and lifting) has been considered. The target task is the same as in [33], to have a fair comparison between the centralized and decentralized approaches. In more detail, the task consists of the robot interacting with the object (a 6 cm cube) placed on a table. The goal is to successfully grasp and lift the cube off the table. The task can be split into several phases that need to be completed in a sequence in order to have a successful execution of the task. The phases are:

- Reaching: the agents receive a reward if the robot's end-effector approaches the cube;
- Grasping: a grasping reward is given if both of the fingers of the gripper touch the cube, so that the grasping phase is fulfilled;
- Lifting: when the cube is lifted off the surface, the lifting reward is provided, and this constitutes a successful episode.

The complete task can be thought of as accomplishing the sub-tasks that lead to achieving the main objective.

3. Methodology

3.1. Preliminaries

In this paper, the RL problem is modeled as a discrete time-continuous Markov decision process (MDP). Markov decision processes are the mathematical foundation for RL [3,34]. A Markov decision process is a tuple (S, P, R, A), where S is a state set, P is the transition probability matrix, R is a reward function of the agent based on the current state and the chosen action, and A is a finite set of actions available to the agent. The objective of the agent is to learn a policy π that maximizes the expected return, whilst achieving the maximum cumulative reward. A policy π represents the description of behaviors of an agent. A stationary policy $\pi : S \rightarrow A$ is a probability distribution over actions to be taken in each state. A deterministic policy has a probability of one for some action in a state.

Reinfocement learning methods can be divided into three main categories:

- Value-based methods, which learn a value function that is then used to derive a policy;
- Policy-based methods, which have a parameterized policy and directly search for the optimal policy;
- Actor-critic methods, which learn both the value function and the policy. The role of the *critic* is to judge the actions taken by the *actor* on the basis of the learned value function. The actor then updates its policy parameters based on the critic's feedback [35].

In this paper, the actor-critic method is used to learn both the policy and the value function. Based on the way the algorithms generate and use experience data, a further classification can be made into *On-policy* and *Off-policy* algorithms. On-policy methods evaluate the same policy that is used to select actions; whereas, off-policy methods learn the target policy that is different from the behavioral policy used to generate experience. In this paper, two main algorithms are used, the on-policy PPO algorithm and the off-policy SAC algorithm. Further information on these algorithms can be found in [36,37].

3.2. Modular Framework for Decentralized Learning

The multi-agent RL method (MARL) consists of N policies that operate on the state and actions of corresponding agents, as illustrated in Figure 2. Instead of having a single policy with the full observation and action space, multi-agent reinforcement learning (MARL) suggests an explicit split in the state and action space according to the agents (e.g., robots or end-effectors), which allows efficient low-level policy training as well as flexible skill composition [31]. The agents in the system can have information about the whole system (all the states) or just one part of the system (the states that the agent directly influences).

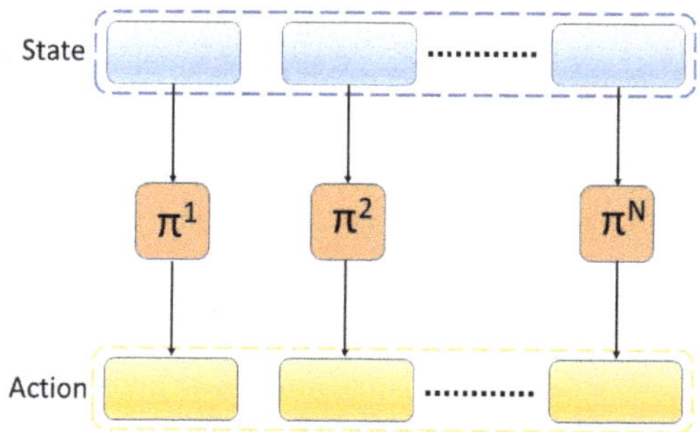

Figure 2. Multi-agent reinforcement learning framework.

In this paper, a multi-agent RL method is proposed, where each of the N policies takes as input the observation of the whole system and outputs an action for the corresponding agent. Without a loss of generality, N = 2 policies are considered in this paper.

All policies share the global critic learned from a single task reward, as mentioned in [38]. The multi-agent system can be composed of an additional component, which is the meta-policy, as illustrated in Figure 3. This framework is comprised of a set of N agents $\pi_1, \ldots \pi_N$ and a meta-policy π_{meta}. In this configuration, the action of two low-level agents (PPO or SAC agents) learn to control the motors, each having an output of four (the control of the first or second part of the robot), and the meta-agent that oversees the process, gives importance to one of the agents. The meta-agent represents a high-level policy that decides which low-level policy to execute among a set of available policies. The low-level agents can first learn the primitive skills separately, which are then combined later on by a meta-agent to execute a complex task. In this paper, the low-level agents have not learned any specific primitive skills, such as positioning, orienting, or grasping. The learning process for both the low-level agents and the meta-agent is executed jointly. This choice is made in order to have comparable results with the single-agent system. In future work, the agent-specific tasks and rewards can first be learned by the primitive agents and later coordinated temporally by the meta-agent.

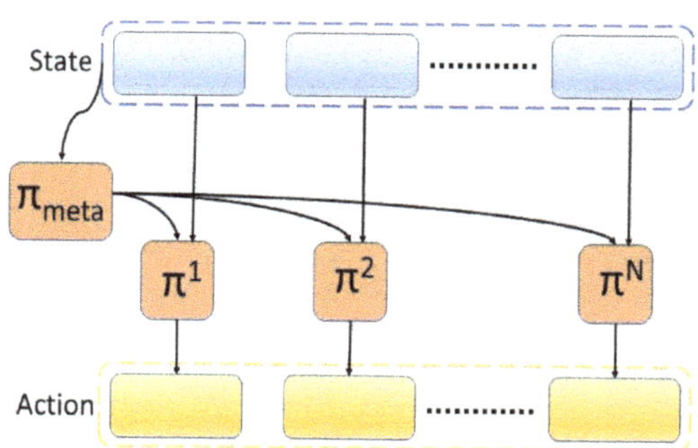

Figure 3. Multi-agent reinforcement learning framework with a meta-agent.

4. Implementation of Proposed Approach

Reinforcement learning requires an interaction between the agent and the environment. In order for this interaction to be successful and for learning to be possible, the environment and task need to be completely defined. The environment is everything outside the agent.

In the case of a robotic manipulator, the environment consists of the robot, the object that needs to be picked up, the table on which the object rests, and the full arena where the robot is placed. The other part that needs to be defined is the task. The task, in this case, is a pick-up task. The rewards that are given to the agent are associated with this task. This section presents a detailed description of the learning environment, the training details, and the network architectures. The final part of this section is used to explain the design of the reward function. This paper is based on the use of a MuJoCo multiphysics simulation engine [39] to model the environment and Pytorch library to train the agent.

4.1. Learning Environment

The learning environment was designed in MuJoCo physics engine in order to simulate the physical system. MuJoCo is a model-based optimization platform that allows fast and accurate simulations, including contact geometries. In order to speed up the simulation, MuJoCo allows the simulation without rendering. A comparison of simulation performance for multiple physics engines is presented in [40], showing that MuJoCo is the fastest environment for robotics-related tasks. The simulation of the robot being used is taken from [41]. With the combination of different models in the simulation, the simulation setup achieved is shown in Figure 4.

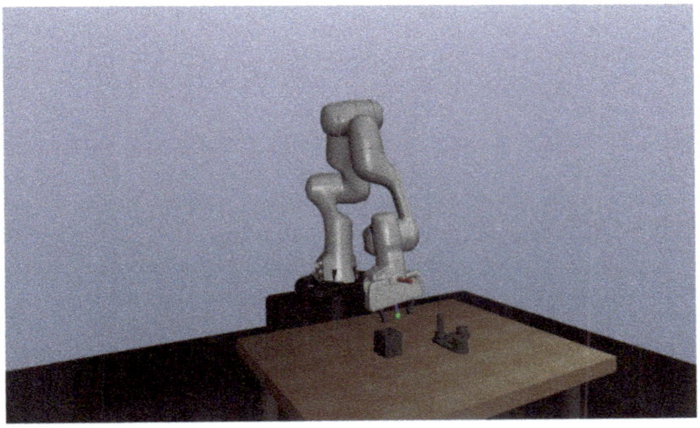

Figure 4. Franka EMIKA panda robot employed in the target grasping task in the MuJoCo environment.

4.2. State-Space

The considered state space is continuous and consists of two main input modalities, the robot information and the target object information:

- Robot information: the data about the robot and its state is a 36-dimensional vector that contains the joint positions, velocities, gripper position, and the end-effector position, velocity, and orientation. The position of the joints is divided and represented by trigonometric functions, i.e., by sine and cosine representations. The robot has 7 DoFs, which gives a vector of 14 values for the position of each joint represented by a sin and cos function. The gripper is defined by a 2-dimensional vector that specifies the position of the left and right fingers. The joint velocities are represented by a single value for each joint, making a 7-dimensional vector. The end-effector data is represented by a 7-dimensional vector, where the position is represented in cartesian coordinates followed by a quaternion to represent the orientation. The end-effector's state also includes a 6-dimensional vector specifying the linear and angular velocity.
- Object information: the data of the object that needs to be picked up is represented by the position and orientation in the world reference frame. In order to facilitate the reaching of the cube, a relative position vector between the object and the robot's gripper is added. The object state, thus, makes a 10-dimensional vector.

4.3. Action Space

The action space consists of an 8-dimensional vector, which represents the 7 joints of the panda robot and the gripper. The last dimension of the action space represents the actuation control of the gripper and is mirrored to have a symmetrical action for the fingers. The action space is split into two equal parts. One part of the manipulator corresponds to the motors of the first 4 joints, while the second part corresponds to the last 3 joints and the gripper's actuation. Dividing the action space in such a way allows for the decentralization of the system and shows the modularity of the robot. Joint velocities coming from the agents are fed into a controller that converts the desired joint velocities into joint torques. The robot is considered to be equipped with torque control compensating for the manipulator dynamics [42]:

$$\boldsymbol{\tau} = \mathbf{B}(\mathbf{q})\boldsymbol{\tau}_{learn} + \boldsymbol{\tau}_f(\dot{\mathbf{q}}) + \mathbf{C}(\mathbf{q}, \dot{\mathbf{q}}) + \mathbf{g}(\mathbf{q}). \quad (1)$$

In this equation, $\mathbf{B}(\mathbf{q})$ represents the robot inertia matrix, $\mathbf{C}(\mathbf{q}, \dot{\mathbf{q}})$ is the robot Coriolis vector, $\mathbf{g}(\mathbf{q})$ is the robot gravitational vector, $\boldsymbol{\tau}_f(\dot{\mathbf{q}})$ is the robot joint friction vector, $\mathbf{q}$ is the current joint position vector, and $\boldsymbol{\tau}$ is the control torque vector. $\boldsymbol{\tau}_{learn}$ is the learned task-related control torque vector mapping the reference joint velocities to joint torques for the target task execution.

$$\boldsymbol{\tau}_{learn} = \mathbf{k}_v(\dot{\mathbf{q}}^d - \dot{\mathbf{q}}), \quad (2)$$

where $\dot{\mathbf{q}}^d$ and $\dot{\mathbf{q}}$ are the reference (i.e., the objective of the learning) and current joint velocities respectively, and:

$$\mathbf{k}_v = diag(8, 7, 6, 4, 2, 0.5, 0.1). \quad (3)$$

$\mathbf{k}_v$ is a diagonal matrix of fixed proportional gains for joint trajectory-tracking purposes. These control gains need to be tuned to allow the robot to track the reference joint velocities $\dot{\mathbf{q}}$. The values that are selected are based on the variable impedance control [43], which results in accurate trajectory tracking performance. If needed, such control gains can be either learned or experimentally tuned to improve trajectory tracking performance.

4.4. Training Details

The algorithms described in Section 3 are used to train the target grasping task. The episodes are divided into 600 time-steps where each of them corresponds to roughly 0.1 s of real-time. At every step of the episode, the actions to be executed in the environment are sampled from Gaussian distributions parameterized by the mean and standard deviation for each action dimension. The agents use a replay buffer that stores the transition experience of the episodes. At every update iteration, random samples of a batch size of 512 are collected from the buffer to update the network parameters. While using a PPO agent, the buffer is cleared and the samples are discarded after performing each update because this is an on-policy algorithm. Both agents have the same parameters, the same state-space is fed into both of the agents, and both agents follow a given policy where the output of each agent is a vector of four values, each corresponding to the given motors. Each agent gives half of the output needed for the system and these outputs are combined and fed into the system. The policy network, as well as the value network, are encoded by MLP—multi-layer perceptron. The policy network consists of three layers and Rectified Linear Unit (ReLU) non-linearity. The input layer, representing the state of the system is 46-dim. The hidden layer size is of 64-dim and the output layer is 8-dim. The output layer of a policy network is two-fold, producing the mean and standard deviation (parametrizing the Gaussian distribution) for each action dimension. The value network outputs only a scalar value, which specifies the corresponding value function of the state. All inputs to the policy and value networks are normalized with running estimates of the mean and variance. Both the policy and value networks are shown in Figure 5.

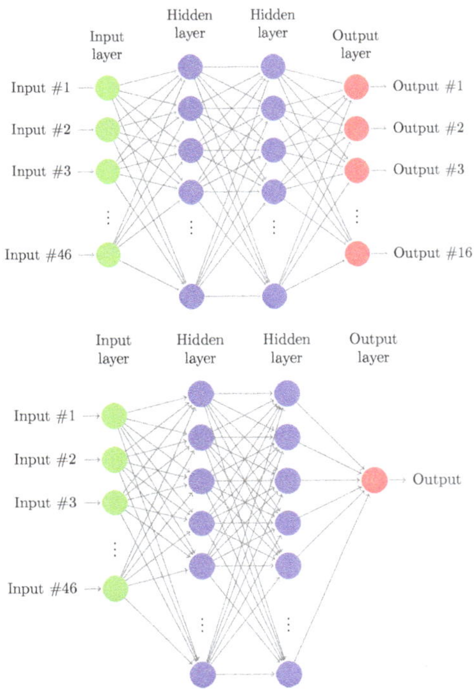

Figure 5. Schematic representation of: (**top**) policy network with 46-dim input (environment states) and 8-dim output (actions); (**bottom**) value network with considered 46-dim input and 1-dim output.

4.5. Reward Shaping

The reward function helps the agents determine if their actions were desirable or not, thus allowing the agents to decide what they need to do. The dense reward signal guides the agent's exploration process and enables efficient learning; however, shaping a reward function is an exhaustive process. Another approach to overcome the exploration issue is to provide an interactive feedback to the agent where the external teacher may overwrite the actions prescribed by the policy [44]. This approach is termed policy shaping, and allows the agent to learn even in the presence of binary reward signals. In this work, the reward shaping approach has been used to overcome the exploration issue for multi-agent learners in order to compare the results with a centralized single-agent that uses the same reward function. Shaping the multi-agent policies using interactive feedback, either from external human teachers or artificial trainers, is a promising direction for future work.

The reward function is divided into two phases, and a separate reward is given for each phase. The reward for the reaching phase is as follows:

$$r_{t,reach} = w_{dist} r_{dist} + w_{vel} r_{vel} + w_{grip} r_{grip}. \tag{4}$$

In the above equation three different rewards are represented:

- r_{dist}: the distance reward which is computed using the relative position of the gripper w.r.t. the object;
- r_{vel}: the velocity reward that takes into account the end-effector velocity vector when approaching the object;
- r_{grip}: the gripper open reward depending upon the action of the gripper.

The combination of r_{vel} and r_{grip} incentivizes the agents to approach the object with an open gripper and a small velocity. All three contributions (r_{dist}, r_{vel}, r_{grip}) are weighted with respective weights w_{dist} = 0.6, w_{vel} = 0.3, and w_{grip} = 0.1, and defined as:

$$r_{dist} = 1 - tanh(\mathbf{p}_{grip} - \mathbf{p}_{cube}),$$
$$r_{vel} = \begin{cases} 1 - tanh(abs(\mathbf{v}_{ee}/N_{dim}) & if \;\; r_{dist} < 8\text{ cm} \\ 0 & otherwise, \end{cases} \tag{5}$$
$$r_{grip} = \begin{cases} abs(action_{gripper}) & if \;\; action_{gripper} < 0 \\ 0 & otherwise, \end{cases}$$

where $\mathbf{p}_{grip}$ and $\mathbf{p}_{cube}$ denote the position vector of the gripper and the cube in the world reference frame; $\mathbf{v}_{ee}$ represents the velocity of the end-effector, which contains the linear and angular components of the velocity with $N_{dim} = 6$; $action_{gripper}$ specifies the action of the gripper with negative values indicating that the gripper is opening.

The lifting phase of the reward is initiated when the robot's gripper is in close proximity to the object. The distance threshold for switching between the grip and lift rewards is set to 2.5 cm. This is because the nominal cube measures 6 cm, and the robot should be able to grasp the cube when its grip site is within the cube boundaries. The reward for the lifting phase is defined as:

$$r_{t,lift} = r_{dist} + r_{vel} + w_{grip} r_{grip} + r_{grasp} + r_{success}, \tag{6}$$

and consists of the following five contributions: r_{dist} and r_{vel} are the same as the previous equation; r_{gip} reward is now reversed so it gives a reward when the gripper closes in order to grasp the cube. This reward is weighted by w_{grip} = 0.1. Additional rewards are r_{grasp} and $r_{success}$. r_{grasp} is given if both fingers of the gripper make contact with the walls of the cube and $r_{success}$ is an additional reward if the episode is successful, i.e., the task is completed. A successful episode constitutes if the gripper holds the cube at a certain height above the table. r_{grip}; r_{grasp}; $r_{success}$ in the lifting case are defined as:

$$r_{grip} = \begin{cases} abs(action_{gripper}) & if \;\; action_{gripper} > 0 \\ 0 & otherwise, \end{cases}$$
$$r_{grasp} = \begin{cases} 0.5 & if \;\; fingers \;\; in \;\; contact \\ 0 & otherwise, \end{cases} \tag{7}$$
$$r_{success} = \begin{cases} 2 & if \;\; z_{cube} > 0.1 + z_{cube_{initial}} \\ 0 & otherwise. \end{cases}$$

Different iterations were performed to set the values of the grasping reward r_{grasp} and the success reward $r_{success}$. In particular, setting higher values led to a very high variance in the reward, causing training inefficiency. The selected values encourage the learning agent to accumulate more reward in the lifting phase by accomplishing grasping and lifting of the cube, while avoiding too high of a variance in training results. The total reward r_t is then computed as the accumulated sum of $r_{t,reach}$ in Equation (4) and $r_{t,lift}$ in Equation (6).

$$r_t = r_{t,reach} + r_{t,lift}. \tag{8}$$

4.6. Software Implementation

The software implementation is available at the following GitHub repository [45].

5. Results

In this section, the algorithms used for the training are evaluated and compared with the centralized approach. The reward function presented in Section 4.5 has been shaped and optimized for the single-agent control. For the decentralized and multi-agent control, the same reward function is used to provide a proper comparison with the centralized approach. The following multi-agent algorithms are considered:

1. One meta-PPO agent and two low level SAC agents;
2. Two PPO agents;
3. Two SAC agents;
4. One PPO agent and one SAC agent.

In the first configuration, the two low-level agents learned to control the first or second part of the robot (each having an output of four), and the meta-agent oversees the process and gives importance to one of the agents. The other three configurations involve two separate agents, where each one controls one part of the action space, i.e., each controls four motors of the robot. The two agents do not communicate with each other, and each one strives to make the optimal decision knowing the observation space, but not knowing the actions of the other agent. The two agents follow either the PPO policy, the SAC policy, or the combination of both.

A final configuration is considered where only one agent is used, but the action space remains divided. Having the action space divided in such a way allows the agent to look at the actions it takes as two separate vectors. The output of the Policy network is thus two separate four-element vectors, each corresponding to the first and last four motors of the robot. In this way, only one agent is used, which looks at the action space in a decentralized way. The algorithms considered for decentralized single-agent control are the following:

- One SAC agent;
- One PPO agent.

A video showing the progress of the training employing the decentalized single-agent SAC controller is available at the GitHub link [45]. In the video, the improvements in the learning of the target task (i.e., grasping and lifting of the part) are shown, highlighting the capabilities of the method.

5.1. Decentralized Multi-Agent Approach

5.1.1. Meta-PPO Agent with Two Low Level SAC Agents

This type of baseline is a hierarchical framework that composes of a meta-policy that coordinates the two low-level agents, which control the robot joints following the SAC policy. In Figure 6, results are shown for the behavior of this algorithm. At the beginning of the training, a steady and relatively fast increase in the rewards is seen; this can be explained by the fact that everything in the experience is new to the robot, and the functional algorithms make the proper use of this experience for learning. However, after approximately three million time steps, the rewards plummet, and the agents enter a local minimum, which they do not escape, regardless of the number of iterations. The meta-agent's interaction with the low-level agents comes to a point where it thinks the maximum successful reward has been achieved. This training has been run for 10 million time steps, and the reward stays in the same range of 50–100, without a successful episode.

Figure 6. Progression of mean accumulated reward for the meta-agent with two low-level SAC agents. Training results are reported for three experiments with different random seeds. The standard deviation of three runs is shown in the plot.

5.1.2. Two PPO Agents

The results for the case of two PPO agents are shown in Figure 7. As expected, the PPO agent has a more significant non-monotonic behavior. This experiment shows that a high reward of around 500 is achieved near the two million time steps, whereas the same algorithm, when used in the centralized single-agent setting, reaches the maximum reward of 250 at two million steps [46].

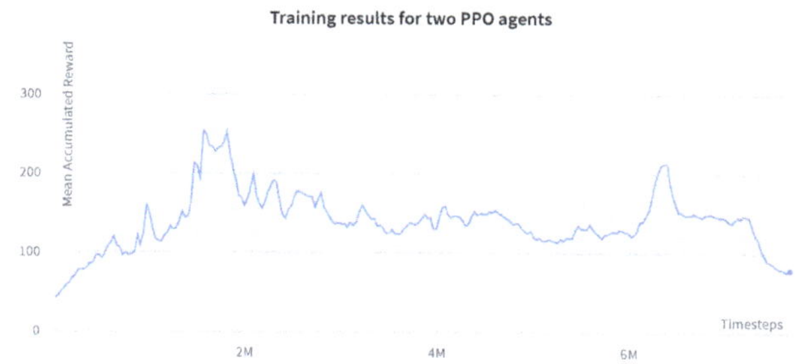

Figure 7. Progression of mean accumulated reward for the two PPO agents.Training results are reported for four experiments with different random seeds. The standard deviation of four runs is shown in the plot.

At certain times before the two million steps, very high rewards can be seen. At these points, the robot manages to grasp the cube and the learning agent transitions to the second phase of the reward described in Section 4.5. However, unfortunately, it does not result in the lifting of the cube. This can be explained by the fact that the two agents prioritize the actions coming from their own policies instead of cooperating towards a common goal.

5.1.3. Two SAC Agents

In this configuration, two SAC agents are used in the multi-agent setting. The mean accumulated reward is plotted in Figure 8. A very similar conclusion can be inferred from the two SAC agents multi-agent controller. A clear rise in the total reward obtained can be read from the graph with a much more robust and monotonic increase, as expected to obtain an off-policy algorithm.

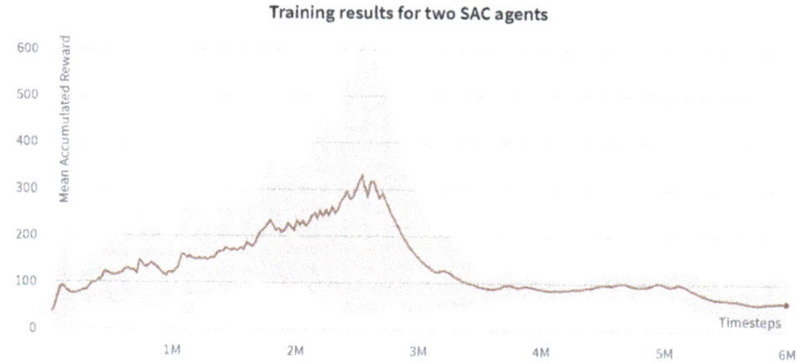

Figure 8. Progression of mean accumulated reward for the two SAC agents. Training results are reported for four experiments with different random seeds. The standard deviation of four runs is shown in the plot.

5.1.4. One SAC Agent One PPO Agent

In Figure 9, the training results for the combination of an SAC and PPO agent are shown. Using two separate algorithms, which follow two separate policies does not yield promising results. There is a slight uptrend in the reward at the beginning but the whole curve is relatively flat in the region of 50–150 for the duration of the whole run, which lasted for eight million timesteps.

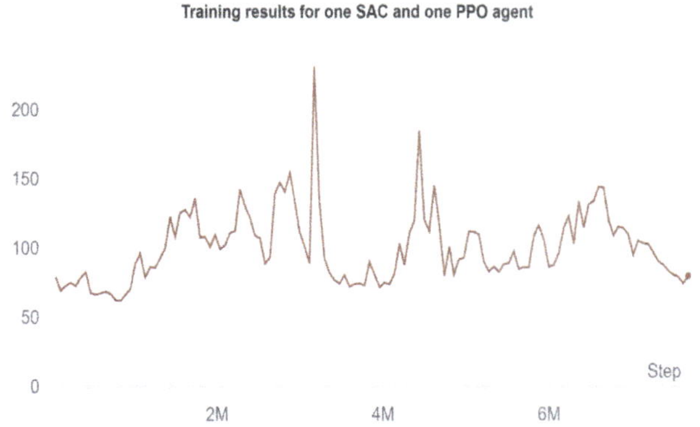

Figure 9. Progression of the accumulated reward for one SAC and one PPO agent.

5.2. Decentralized Single-Agent

5.2.1. Decentralized Single-Agent SAC

When using only one SAC agent with a divided action space, the following results achieved are shown in Figures 10 and 11.

Figure 10. Progression of the accumulated reward for the decentralized single-agent SAC.

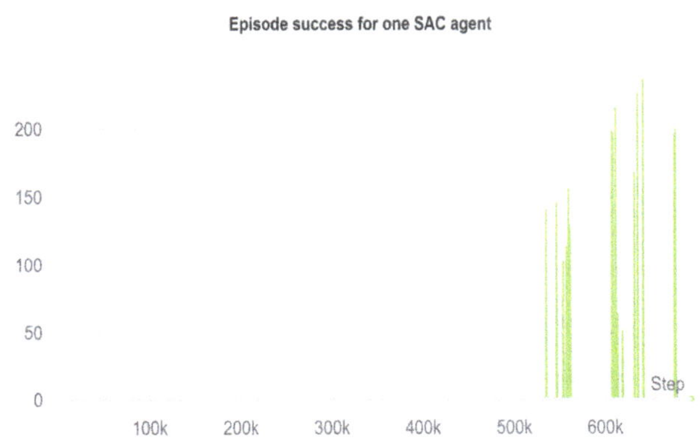

Figure 11. Episode success rate for the decentralized single-agent SAC.

The figure shows a steady uptrend of the reward. At 400K timesteps, the robot learns to grasp the cube, while at 600K, the training results in a successful lifting of the cube. Comparing these results with the single-agent centralized approach in [46], the iterations needed for successful task completion are halved, achieving a reward of 1200 at 600K timesteps, leading to a much faster and efficient learning of the task. The division of the action space in single-agent helps to save the memory, and the training efficiency improves due to the fact that a common learning goal is shared through the same reward function, while the actions are redirected to the corresponding motors in a decentralized manner for the execution.

5.2.2. Decentralized Single-Agent PPO

In this configuration, a single PPO agent is used with a decentralized action space and the results are shown in Figure 12.

Figure 12. Progression of the mean accumulated reward for the decentralized single-agent PPO. Training results are reported for three experiments with different random seeds. The standard deviation of three runs is shown in the plot.

The comparison of the results achieved through decentralized single-agent SAC and PPO is shown in Figure 13. It is shown that the decentralized PPO agent required more iterations in order to reach a successful episode in comparison to the SAC agent. The PPO agent also tends to stagnate at certain areas, which can be seen after the two million steps where the agent is performing positively, but later falls to a local minimum.

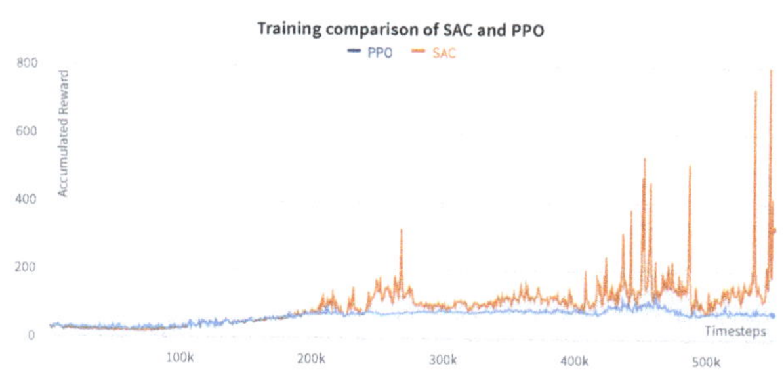

Figure 13. Comparison of accumulated reward for the decentralized single-agent SAC and PPO.

5.3. Comparison of Centralized and Decentralized Approach

The training comparison between the centralized and decentralized approach for PPO and SAC agents is shown in Figures 14 and 15. The decentralized PPO algorithm accumulates higher average reward than the centralized PPO; however, it also shows a non-monotonic behavior. Whereas, the learning curve of the centralized PPO agent is smooth, as demonstrated in Figure 14.

Figure 14. Comparison of accumulated reward for the centralized and decentralized PPO agents.

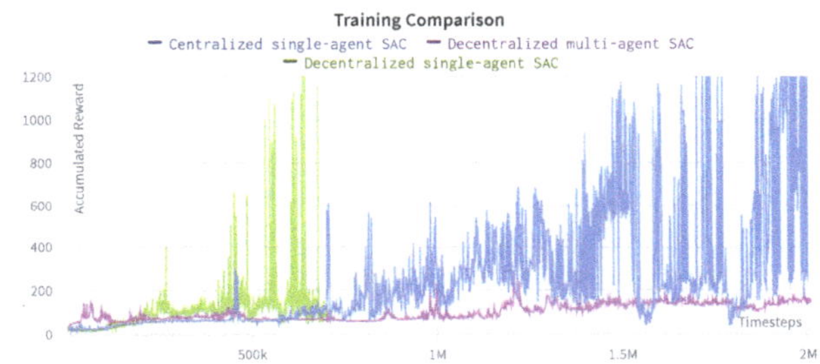

Figure 15. Comparison of accumulated reward for the centralized and decentralized SAC agents.

On the other hand, the decentralized single-agent SAC in Figure 15 demonstrates a much faster and efficient learning process as it accumulates a very high average reward and shows the task success in just 600K timesteps while the centralized SAC agent needs at least 2M steps to reach success. The multi-agent SAC curve instead is relatively flat and does not reach any task success in the first 2M timesteps.

Table 1 summarizes the results of all analyzed approaches and shows the comparison of maximum accumulated reward achieved in an episode considering the first 2M steps of training. The decentralized single-agent SAC is the most efficient approach of all, since the agent accumulates the highest reward, while the combination of multi-agent SAC and PPO achieves the lowest reward. This might be due to the lack of coordination among agents in the multi-agent configuration as there are no explicit terms in the reward to encourage such coordination. Therefore, an interesting direction for the future work is to shape the reward function that encourages coordination among agents in a multi-agent setting.

Table 1. Maximum accumulated reward in the first 2M time steps.

Approach	Maximum Accumulated Reward
One meta-PPO agent and two low-level SAC agents	346
Two PPO agents	580
Two SAC agents	607
One PPO agent and one SAC agent	130
Decentralized single SAC agent	1203
Decentralized single PPO agent	480

6. Discussion

Similar results and behaviors can be seen for all multi-agent approaches. The learning efficiency is improved at the beginning of the training before the stagnation of the learning curve. Therefore, the multi-agent division doesn't improve the training efficiency for the completed task. However, it is still better from the computational point of view as the division of the action space provides modularity to the overall system, allowing to train the agents in a distributed manner. Moreover, larger expected rewards can be achieved by the off-policy SAC multi-agent compared to the PPO multi-agent, since it takes advantage of past data storage. The PPO agent is slower than the SAC agent in both the centralized and decentralized approach. Most of the higher rewards achieved during the multi-agent runs can be correlated with the gripper touching the cube with a good approaching velocity. Nevertheless, a lack of coordination and a sparse reward setup for the second phase in charge of lifting the cube prevents it from completing the task. In the reaching phase, a very high reward was achieved by the first agent that was, however, overturned in the lifting phase where the actions of the second agent had more significance. This sort of behavior can be attributed to the fact that the reward is shaped to accommodate the centralized single-agent learning. Since the reward is highly sensitive, and small variations can lead to the difference between a successful episode and an unsuccessful run, the reward function can be augmented with different aspects to accommodate the needs of multi-agent learning as was done for the centralized single-agent. In addition, having a separate reward function for each of the agents and augmenting the agents' input with the information about prior sub-tasks could improve the performance of the decentralized multi-agent learning.

The video available at the GitHub link [45] shows the evolution of the learning process for the decentalized single-agent SAC controller, having the robot able to learn the target task (i.e., grasping and lifting of the target part). The policy of the training run with the highest accumulated reward is selected as the control actions to be applied to the robot to execute the learned task. As in [33], the learned control torques or joint positions/velocities can be transferred to the real robot to perform the given application.

7. Conclusions

In this paper, a decentralized and multi-agent approach has been formulated in the form of an RL problem, demonstrating the possibility of decentralizing a single manipulator controller by applying multiple agents in learning continuous actions. The idea of this paper has been to compare the feasibility of decentralization for a single robot to show the modularity in the joints, as well as the comparison between the centralized and decentralized approach. Results show that it is possible to decentralize the control action on the robot. Using multiple agents has shown the stagnation of the learning process when using the same reward function that is used for a single-agent. It is believed that this is due to the lack of communication between the agents and the generality of the reward when considering two agents. A sparser reward needs to be added to the system and agent specific rewards can be considered.

Future work will be devoted to deeper investigating of the proposed approaches, implementing improved methodologies based on the achieved results. A deeper investigation and analysis of the results (e.g., increasing the number of runs of each method to better highlight local minima issues, etc.) will also be conducted in order to provide further insights. In addition, the reward function will be tuned for each agent separately to improve the performance of the methods. Then, adding a sparse reward function, as well as agent-specific rewards, could result in a successful execution of the task using the decentralized approach. Meta-learning will be deeply investigated in order to further generalize the methodology. The transfer of the learned policies to manipulators (e.g., to the Franka EMIKA panda robot) will be performed to show the applicability of the proposed methods in real tasks.

Author Contributions: Conceptualization, L.R., A.A.S., J.S.V.S., D.P. (Damjan Pecioski) and D.P. (Dario Piga); methodology, L.R., A.A.S., J.S.V.S., D.P. (Damjan Pecioski) and D.P. (Dario Piga); software, J.S.V.S., D.P. (Damjan Pecioski) and A.A.S.; validation, J.S.V.S., D.P. (Damjan Pecioski), A.A.S. and L.R.; formal analysis, J.S.V.S., D.P. (Damjan Pecioski), A.A.S. and L.R.; investigation, L.R., A.A.S. and D.P. (Damjan Pecioski); resources, L.R.; data curation, J.S.V.S., D.P. (Damjan Pecioski) and A.A.S.; writing—original draft preparation, L.R., A.A.S., J.S.V.S., D.P. (Damjan Pecioski) and D.P. (Dario Piga); writing—review and editing, A.A.S. and L.R.; visualization, L.R., A.A.S., J.S.V.S., D.P. (Damjan Pecioski) and D.P. (Dario Piga); supervision, L.R., D.P. (Dario Piga) and F.B.; project administration, L.R.; funding acquisition, L.R. All authors have read and agreed to the published version of the manuscript.

Funding: This research was funded by the project ASSASSINN, funded from H2020 CleanSky 2, grant agreement n. 886977.

Institutional Review Board Statement: Not applicable.

Informed Consent Statement: Not applicable.

Data Availability Statement: The software to replicate the achieved results can be found at the link in [45].

Acknowledgments: The work has been developed within the project ASSASSINN, funded from H2020 CleanSky 2 under grant agreement n. 886977.

Conflicts of Interest: The authors declare no conflict of interest.

References

1. Rajan, K.; Saffiotti, A. Towards a Science of Integrated AI and Robotics. *Artif. Intell.* **2017**, *247*, 1–9. [CrossRef]
2. Van Roy, V.; Vertesy, D.; Damioli, G. AI and robotics innovation. In *Handbook of Labor, Human Resources and Population Economics*; Springer: Berlin/Heidelberg, Germany, 2020; pp. 1–35.
3. Sutton, R.S.; Barto, A.G. *Reinforcement Learning: An Introduction*; MIT Press: Cambridge, MA, USA, 1998; Volume 22447.
4. Mnih, V.; Kavukcuoglu, K.; Silver, D.; Rusu, A.A.; Veness, J.; Bellemare, M.G.; Graves, A.; Riedmiller, M.; Fidjeland, A.K.; Ostrovski, G.; et al. Human-level control through deep reinforcement learning. *Nature* **2015**, *518*, 529–533. [CrossRef] [PubMed]
5. Silver, D.; Huang, A.; Maddison, C.J.; Guez, A.; Sifre, L.; Van Den Driessche, G.; Schrittwieser, J.; Antonoglou, I.; Panneershelvam, V.; Lanctot, M.; et al. Mastering the game of Go with deep neural networks and tree search. *Nature* **2016**, *529*, 484–489. [CrossRef] [PubMed]
6. Kober, J.; Bagnell, J.A.; Peters, J. Reinforcement learning in robotics: A survey. *Int. J. Robot. Res.* **2013**, *32*, 1238–1274. [CrossRef]
7. Cruz, F.; Parisi, G.I.; Wermter, S. Learning contextual affordances with an associative neural architecture. In Proceedings of the 24th European Symposium on Artificial Neural Networks, Bruges, Belgium, 27–29 April 2016; pp. 665–670.
8. Yen-Chen, L.; Zeng, A.; Song, S.; Isola, P.; Lin, T.Y. Learning to See before Learning to Act: Visual Pre-training for Manipulation. *arXiv* **2021**, arXiv:2107.00646.
9. Cruz, F.; Dazeley, R.; Vamplew, P.; Moreira, I. Explainable robotic systems: Understanding goal-driven actions in a reinforcement learning scenario. *Neural Comput. Appl.* **2021**, 1–18. [CrossRef]
10. Cruz, F.; Wüppen, P.; Fazrie, A.; Weber, C.; Wermter, S. Action selection methods in a robotic reinforcement learning scenario. In Proceedings of the 2018 IEEE Latin American Conference on Computational Intelligence (LA-CCI), Gudalajara, Mexico, 7–9 November 2018; pp. 1–6. [CrossRef]
11. Cruz, F.; Wüppen, P.; Magg, S.; Fazrie, A.; Wermter, S. Agent-advising approaches in an interactive reinforcement learning scenario. In Proceedings of the 2017 Joint IEEE International Conference on Development and Learning and Epigenetic Robotics (ICDL-EpiRob), Lisbon, Portugal, 18–21 September 2017; pp. 209–214. [CrossRef]
12. Rahmatizadeh, R.; Abolghasemi, P.; Bölöni, L.; Levine, S. Vision-based multi-task manipulation for inexpensive robots using end-to-end learning from demonstration. In Proceedings of the 2018 IEEE International Conference on Robotics and Automation (ICRA), Brisbane, QLD, Australia, 21–25 May 2018; pp. 3758–3765.
13. Gu, S.; Holly, E.; Lillicrap, T.; Levine, S. Deep reinforcement learning for robotic manipulation with asynchronous off-policy updates. In Proceedings of the 2017 IEEE International Conference on Robotics and Automation (ICRA), Singapore, 29 May–3 June 2017; pp. 3389–3396.
14. Joshi, S.; Kumra, S.; Sahin, F. Robotic grasping using deep reinforcement learning. In Proceedings of the 2020 IEEE 16th International Conference on Automation Science and Engineering (CASE), Hong Kong, China, 20–21 August 2020; pp. 1461–1466.
15. Lillicrap, T.P.; Hunt, J.J.; Pritzel, A.; Heess, N.; Erez, T.; Tassa, Y.; Silver, D.; Wierstra, D. Continuous control with deep reinforcement learning. *arXiv* **2015**, arXiv:1509.02971.
16. Levine, S.; Pastor, P.; Krizhevsky, A.; Ibarz, J.; Quillen, D. Learning hand-eye coordination for robotic grasping with deep learning and large-scale data collection. *Int. J. Robot. Res.* **2018**, *37*, 421–436. [CrossRef]
17. Busoniu, L.; De Schutter, B.; Babuska, R. Decentralized reinforcement learning control of a robotic manipulator. In Proceedings of the 2006 9th International Conference on Control, Automation, Robotics and Vision, Singapore, 5–8 December 2006; pp. 1–6.
18. Leottau, D.L.; Ruiz-del Solar, J.; Babuška, R. Decentralized reinforcement learning of robot behaviors. *Artif. Intell.* **2018**, *256*, 130–159. [CrossRef]
19. Panait, L.; Luke, S. Cooperative multi-agent learning: The state of the art. *Auton. Agents Multi-Agent Syst.* **2005**, *11*, 387–434. [CrossRef]
20. Andrychowicz, M.; Wolski, F.; Ray, A.; Schneider, J.; Fong, R.; Welinder, P.; McGrew, B.; Tobin, J.; Abbeel, P.; Zaremba, W. Hindsight experience replay. *arXiv* **2017**, arXiv:1707.01495.

21. Lauer, M.; Riedmiller, M. An algorithm for distributed reinforcement learning in cooperative multi-agent systems. In Proceedings of the Seventeenth International Conference on Machine Learning, Stanford, CA, USA, 29 June–2 July 2000.
22. Kapetanakis, S.; Kudenko, D. Reinforcement learning of coordination in heterogeneous cooperative multi-agent systems. In *Adaptive Agents and Multi-Agent Systems II*; Springer: Berlin/Heidelberg, Germany, 2004; pp. 119–131.
23. Matignon, L.; Laurent, G.J.; Le Fort-Piat, N. Hysteretic q-learning: An algorithm for decentralized reinforcement learning in cooperative multi-agent teams. In Proceedings of the 2007 IEEE/RSJ International Conference on Intelligent Robots and Systems, San Diego, CA, USA, 29 October–2 November 2007; pp. 64–69.
24. Tuyls, K.; Hoen, P.J.; Vanschoenwinkel, B. An evolutionary dynamical analysis of multi-agent learning in iterated games. *Auton. Agents Multi-Agent Syst.* **2006**, *12*, 115–153. [CrossRef]
25. Laurent, G.J.; Matignon, L.; Fort-Piat, L. The world of independent learners is not Markovian. *Int. J. Knowl.-Based Intell. Eng. Syst.* **2011**, *15*, 55–64. [CrossRef]
26. Vidhate, D.; Kulkarni, P. Cooperative machine learning with information fusion for dynamic decision making in diagnostic applications. In Proceedings of the 2012 International Conference on Advances in Mobile Network, Communication and Its Applications, Bangalore, India, 1–2 August 2012; pp. 70–74.
27. Theodorou, E.; Buchli, J.; Schaal, S. Reinforcement learning of motor skills in high dimensions: A path integral approach. In Proceedings of the 2010 IEEE International Conference on Robotics and Automation, Anchorage, AK, USA, 3–7 May 2010; pp. 2397–2403.
28. Kuba, J.G.; Wen, M.; Yang, Y.; Meng, L.; Gu, S.; Zhang, H.; Mguni, D.H.; Wang, J. Settling the Variance of Multi-Agent Policy Gradients. *arXiv* **2021**, arXiv:2108.08612.
29. Schwager, M.; Rus, D.; Slotine, J.J. Decentralized, adaptive coverage control for networked robots. *Int. J. Robot. Res.* **2009**, *28*, 357–375. [CrossRef]
30. Sartoretti, G.; Paivine, W.; Shi, Y.; Wu, Y.; Choset, H. Distributed learning of decentralized control policies for articulated mobile robots. *IEEE Trans. Robot.* **2019**, *35*, 1109–1122. [CrossRef]
31. Lee, Y.; Yang, J.; Lim, J.J. Learning to coordinate manipulation skills via skill behavior diversification. In Proceedings of the International Conference on Learning Representations, Addis Ababa, Ethiopia, 26–30 April 2020.
32. Ha, H.; Xu, J.; Song, S. Learning a decentralized multi-arm motion planner. *arXiv* **2020**, arXiv:2011.02608.
33. Shahid, A.A.; Roveda, L.; Piga, D.; Braghin, F. Learning continuous control actions for robotic grasping with reinforcement learning. In Proceedings of the 2020 IEEE International Conference on Systems, Man, and Cybernetics (SMC), Toronto, ON, Canada, 11–14 October 2020; pp. 4066–4072.
34. Littman, M.L. Value-function reinforcement learning in Markov games. *Cogn. Syst. Res.* **2001**, *2*, 55–66. [CrossRef]
35. Lazaric, A.; Restelli, M.; Bonarini, A. Reinforcement learning in continuous action spaces through sequential monte carlo methods. *Adv. Neural Inf. Process. Syst.* **2007**, *20*, 833–840.
36. Schulman, J.; Wolski, F.; Dhariwal, P.; Radford, A.; Klimov, O. Proximal policy optimization algorithms. *arXiv* **2017**, arXiv:1707.06347.
37. Haarnoja, T.; Zhou, A.; Hartikainen, K.; Tucker, G.; Ha, S.; Tan, J.; Kumar, V.; Zhu, H.; Gupta, A.; Abbeel, P.; et al. Soft actor-critic algorithms and applications. *arXiv* **2018**, arXiv:1812.05905.
38. Lowe, R.; Wu, Y.; Tamar, A.; Harb, J.; Abbeel, P.; Mordatch, I. Multi-agent actor-critic for mixed cooperative-competitive environments. *arXiv* **2017**, arXiv:1706.02275.
39. Todorov, E.; Erez, T.; Tassa, Y. Mujoco: A physics engine for model-based control. In Proceedings of the 2012 IEEE/RSJ International Conference on Intelligent Robots and Systems, Vilamoura-Algarve, Portugal, 7–12 October 2012; pp. 5026–5033.
40. Erez, T.; Tassa, Y.; Todorov, E. Simulation tools for model-based robotics: Comparison of bullet, havok, mujoco, ode and physx. In Proceedings of the 2015 IEEE International Conference on Robotics and Automation (ICRA), Seattle, WA, USA, 26–30 May 2015; pp. 4397–4404.
41. Zhu, Y.; Wong, J.; Mandlekar, A.; Martín-Martín, R. robosuite: A Modular Simulation Framework and Benchmark for Robot Learning. *arXiv* **2020**, arXiv:2009.12293.
42. Massa, D.; Callegari, M.; Cristalli, C. Manual guidance for industrial robot programming. *Ind. Robot Int. J.* **2015**, *42*, 457—465. [CrossRef]
43. Martín-Martín, R.; Lee, M.A.; Gardner, R.; Savarese, S.; Bohg, J.; Garg, A. Variable impedance control in end-effector space: An action space for reinforcement learning in contact-rich tasks. In Proceedings of the 2019 IEEE/RSJ International Conference on Intelligent Robots and Systems (IROS), Macau, China, 3–8 November 2019; pp. 1010–1017.
44. Moreira, I.; Rivas, J.; Cruz, F.; Dazeley, R.; Ayala, A.; Fernandes, B. Deep Reinforcement Learning with Interactive Feedback in a Human–Robot Environment. *Appl. Sci.* **2020**, *10*, 5574. [CrossRef]
45. Sesin, J.S.V.; Pecioski, D. GitHub Repository: Software for Decentralized and Multi Agent Control of Franka Emika Panda Robot. Available online: https://github.com/jvidals09/Decentralized-and-multi-agent-control-of-Franka-Emika-Panda-robot-in-continuous-task-execution (accessed on 29 September 2021).
46. Shahid, A.A. Continuous Control Actions Learning with Performance Specifications through Reinforcement Learning. Master's Thesis, Politecnico di Milano, Milan, Italy, 2020. Available online: http://hdl.handle.net/10589/164660 (accessed on 3 October 2021).

applied sciences

Article

Predictive Control for Small Unmanned Ground Vehicles via a Multi-Dimensional Taylor Network

Yuzhan Wu [1], Chenlong Li [2,*], Changshun Yuan [2], Meng Li [2,3] and Hao Li [4]

1 School of Automation Science and Electrical Engineering, Beihang University, Beijing 100191, China; wyzbuaa@163.com
2 Hangzhou Innovation Institute, Beihang University, Hangzhou 310051, China; yuanchang61@buaa.edu.cn (C.Y.); limeng2@sztu.edu.cn (M.L.)
3 College of Big Data and Internet, Shenzhen Technology University, Shenzhen 518118, China
4 School of Automation Engineering, University of Electronic Science and Technology of China, Chengdu 611731, China; haoli_uestc@163.com
* Correspondence: lichenlong007@126.com

Abstract: Tracking control of Small Unmanned Ground Vehicles (SUGVs) is easily affected by the nonlinearity and time-varying characteristics. An improved predictive control scheme based on the multi-dimensional Taylor network (MTN) is proposed for tracking control of SUGVs. First, a MTN model is used as a predictive model to construct a SUGV model and back propagation (BP) is taken as its learning algorithm. Second, the predictive control law is designed and the traditional objective function is improved to obtain a predictive objective function with a differential term. The optimal control quantity is given in real time through iterative optimization. Meanwhile, the stability of the closed-loop system is proved by the Lyapunov stability theorem. Finally, a tracking control experiment on the SUGV model is used to verify the effectiveness of the proposed scheme. For comparison, traditional MTN and Radial Basis Function (RBF) predictive control schemes are introduced. Moreover, a noise disturbance is considered. Experimental results show that the proposed scheme is effective, which ensures that the vehicle can quickly and accurately track the desired yaw velocity signal with good real-time, robustness, and convergence performance, and is superior to other comparison schemes.

Keywords: multi-dimensional Taylor network; predictive control; nonlinear system; SUGV; predictive model

Citation: Wu, Y.; Li, C.; Yuan, C.; Li, M.; Li, H. Predictive Control for Small Unmanned Ground Vehicles via a Multi-Dimensional Taylor Network. *Appl. Sci.* **2022**, *12*, 682. https://doi.org/10.3390/app12020682

Academic Editor: Emanuele Carpanzano

Received: 16 December 2021
Accepted: 8 January 2022
Published: 11 January 2022

1. Introduction

With the rapid development of artificial intelligence, big data, and data fusion technology, car driving tends to be intelligent and unmanned. Recently, research on Unmanned Ground Vehicles (UGVs) has become a hot topic at home and abroad [1–3]. Its core technologies include environmental awareness, vehicle positioning, path planning, and path tracking control [4–7]. Path tracking control, as a key core issue of unmanned driving technology, has attracted much attention from scholars and it ensures that the car drives on the specified path [7–9]. So, research on path tracking control has both theoretical and practical significance.

However, UGV path tracking control is a strongly nonlinear process, which contains time-varying characteristics, uncertainty, etc., and a precise mathematical model is difficult to obtain [10]. As artificial intelligence develops, neural networks can solve the above problem well due to their nonlinear approximation capabilities [11]. For example, a new end-to-end autonomous control method was proposed to simplify the separate modules in the traditional control pipeline into a single neural network [12]. A novel application of the biologically inspired computing paradigm was presented for solving the initial value problem (IVP) of electric circuits based on a nonlinear RL model by exploiting the competency of accurate modeling with a feed forward artificial neural network, the global search efficacy of genetic algorithms, and a rapid local search with sequential quadratic programming [13]. Moreover, predictive control, rooted in industrial engineering, is widely used. Usually, the schemes that combine a neural network model and predictive control are more popular and useful to solve the nonlinear control problem. For example, physics-based recurrent neural network modeling approaches were proposed for a general class of nonlinear dynamic process systems to improve the prediction accuracy by incorporating a priori process knowledge, and the proposed physics-based RNN models were utilized in model predictive controllers [14]. A neural-network-based technique for developing nonlinear dynamic models from empirical data for a model predictive control (MPC) algorithm was presented [15]. However, for the neural networks, they easily reach the local minimum [16].

To solve the issues with nonlinearity, time-varying characteristics, uncertainty, real time, etc. and overcome the problem of neural networks, we adopt a multi-dimensional Taylor network (MTN) and collective solutions. The collective solutions are the most effective as they take advantage of each method. The MTN is an inclusive platform that was first proposed by Zhou and Yan and can be used to describe the nonlinear dynamic characteristics of a system without prior knowledge of the structural information on the system [17,18]. It has been successfully applied to system modeling [19,20] and nonlinear control [21–24]. In this paper, an improved predictive control scheme based on a MTN is proposed to realize real-time output tracking control of Small Unmanned Ground Vehicles (SUGVs) relative to a given reference path without state feedback. First, the MTN is used as a predictive model to construct the SUGV model and back propagation (BP) is taken as its learning algorithm. Second, a predictive control law is designed to control the SUGV and the traditional objective function is improved to obtain a predictive objective function with a differential term. The optimal control quantity is given in real time through iterative optimization. Meanwhile, the stability of the closed-loop system is proved according to the Lyapunov stability theorem. Based on that, an adjustment method for MTN control law parameters is obtained using the predictive accuracy of the MTN model and the control coefficient. Finally, a tracking control experiment on the SUGV is used to verify the effectiveness of the proposed scheme. For comparison, traditional MTN and RBF predictive control schemes are introduced.

By the above analysis, the main contributions are as follows:

(1) Although tracking control schemes for SUGVs have been proposed, their schemes have a complex design and are difficult to use in practical applications. Our proposed scheme has a simple structure and is easy to implement.
(2) As the proposed control scheme adapts online learning, no training process is required. Meanwhile, with its simple structure, the strong approximation capability, and the low complexity of the MTN, desirable real-time performance and speed are guaranteed.
(3) For the quadratic performance index with the differential term, the differential term can help the system to act in advance, reduce the overshoot, and shorten the adjustment time.

The rest of the paper is arranged as follows. A literature review is presented in Section 2. The materials and methods are given in Section 3. The MTN predictive control scheme with the differential term is summarized in Section 4. Section 5 presents the simulation results and a discussion. Section 6 presents our conclusions.

2. Literature Review

An effective tracking control scheme is critical to UGV path tracking control. Because UGV path tracking control is a strongly nonlinear process, the control methods for previous linear systems seem to be inadequate [25]. At present, the common control methods include adaptive control, sliding mode control, fuzzy control, neural network control, and predictive control [26–35]. For instance, for formation control of multiple vehicles, the leader–follower error model was built and an adaptive controller was designed to address the uncertain relative distance using the dynamic estimation of the leader–follower distance [26]. A multivariable model reference adaptive control (MRAC) scheme was studied for the automatic carrier-landing control problem of unmanned aerial vehicles with the system dynamics of nonlinearity, multivariable coupling, and parametric uncertainty [27]. A vehicle lateral controller was designed for autonomous vehicles based on higher-order sliding mode control [28]. An automatic steering control strategy for unmanned vehicles based on robust backstepping sliding mode control theory was proposed [29]. For the accuracy and stability of driverless buses, a path-tracking controller based on fuzzy pure pursuit control with a front axle reference (FPPC-FAR) was proposed [30]. An obstacle avoidance strategy consisting of path planning and robust fuzzy output feedback control was proposed for unmanned ground vehicles [31]. A novel adaptive neural network robust lateral motion control method was presented that can maintain the yaw stability of an autonomous vehicle while minimizing the lateral path tracking error at the limits of driving conditions [32]. A learning-based predictive control approach was presented for an autonomous racing car [33]. A nonlinear model predictive control optimization algorithm based on the collocation method was proposed for unmanned vehicle trajectory tracking [34]. A trajectory tracking error-based model was used to design a linear model predictive controller and its control action was combined with feed forward and robust control actions [35].

However, all of the above-mentioned control technologies have their own inherent bottlenecks both in theory and in application. For example, for adaptive control, there are many stability methods to be chosen from when designing the control rate, but most of them require strong assumptions. For fuzzy control, simple fuzzy processing will lead to a reduction in control precision and poor system dynamic quality. For neural network control, the closed-loop system is stable only at a certain point, and the exponential function of neurons will also lead to a large number of computations and poor real-time performance. Sliding mode variable structure control can easily fall into a chattering state, and it also needs to solve for factors such as velocity, inertia, acceleration, and the switching surface when near the sliding mode surface. For predictive control based on a mathematical model, it is difficult to establish an accurate mathematical model, or even to build a model. Therefore, collective solutions were adopted in this study.

The multi-dimensional Taylor network is an inclusive platform and applies to nonlinear control particularly effectively. For example, an adaptive predictive control approach based on the MTN was proposed for the real-time tracking control of single-input single-output nonlinear systems with input time-delay [21]. A control scheme based on the MTN was proposed to achieve the real-time output feedback tracking control of multi-input multi-output non-affine nonlinear time-varying discrete systems relative to the given reference signals with online training [22]. An adaptive control approach based on the MTN was proposed to control nonlinear uncertain time-varying systems with noise disturbances [23]. A predictive control scheme on the basis of the MTN, named MTN predictive compensation control, was proposed for single-input single-output nonlinear systems [24]. However, the above-mentioned studies focus on theoretical research and they do not have an application background. Therefore, an improved predictive control scheme based on the MTN is proposed for the tracking control of SUGVs.

3. Materials and Methods

3.1. System Description of the SUGV

A simplified SUGV model is shown in Figure 1. Two front wheels and two rear wheels are combined into a single front wheel and rear wheel, which are simplified into a two-degrees-of-freedom vehicle model. The following assumptions were made: we ignored the effect of steering and suspension; we kept the vehicle's longitudinal speed constant and only considered the lateral movement along the y axis and the yaw movement around the z axis; we ignored the effects of horizontal and vertical aerodynamics; and we considered the side characteristics of the tire in the analysis of the tire force [36–38].

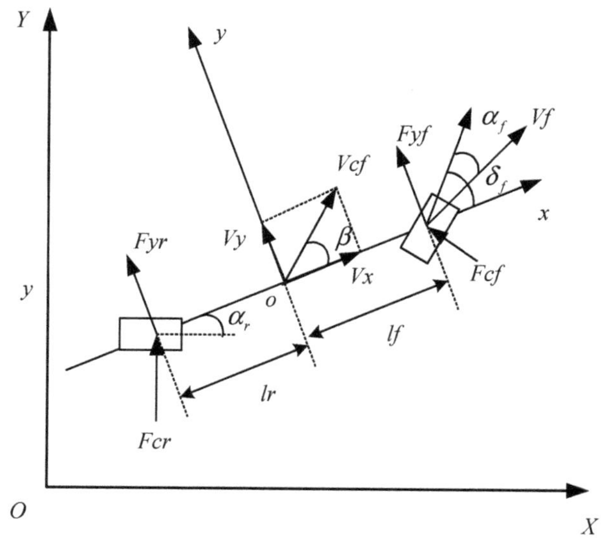

Figure 1. The SUGV model.

The *XOY* coordinate system in Figure 1 is a fixed geodetic coordinate system, and *xoy* is the structure of the coordinate system, which changes with the movement of the body. The force analysis of the above model along the *y* axis and around the *z* axis is obtained.

$$\begin{cases} Ma_y = F_{cf}\cos\delta_f + F_{cr} \\ I_z\dot{w} = I_f F_{cf}\cos\delta_f - I_r F_{cr} \end{cases} \tag{1}$$

where M is the vehicle's weight; I_f is the distance between the center of the front axle and the center of mass of the vehicle; I_r is the distance between the center of the rear axle and the center of mass of the vehicle; I_z is the moment of inertia of the vehicle around the z axis; δ_f is the front wheel angle of the vehicle; w is the yaw velocity of the vehicle; and a_y is the accelerated velocity of the vehicle along the y axis based on the *xoy* coordinate system, which mainly consists of the movement along the y axis and the centripetal acceleration $V_x w$. We can obtain:

$$a_y = \ddot{y} + V_x w \tag{2}$$

where F_{cf} and F_{cr} are the lateral force of the tire applied to the front and rear wheels of the vehicle, respectively. From the defined dynamic model, we can see that the side characteristic of the tire is considered in the analysis of the tire force. When the front wheel angle δ_f and the sideslip angle β are small, the sideslip characteristic of the tire is in a linear range. So, we can obtain

$$F_{cf} = 2C_f\alpha_f \tag{3}$$

$$F_{rf} = 2C_r\alpha_r \tag{4}$$

where C_f and C_r are the cornering stiffnesses of the front and rear tires, respectively. Since there are two front and rear wheels, the force is twice that of a single tire. α_f and α_r are the tire sideslip angle and the angle between the direction of the tire velocity vector and the x axis, respectively. At a small angle, the size of the two angles can be approximately expressed as:

$$\tan\alpha_f \approx \theta < V_f,\ x >= \frac{V_y + l_f\dot{\varphi}}{V_x} \tag{5}$$

$$\tan\alpha_r \approx \theta < V_r,\ x >= \frac{V_y - l_r\dot{\varphi}}{V_x} \tag{6}$$

According to the coordinate system relationship, the side slip angle is $\beta = \tan\beta = \frac{V_y}{V_x}$. From (1)–(6), the differential equation of the vehicle dynamics model is obtained.

$$\begin{cases} \dot{w} = \frac{l_f^2 C_f + l_r^2 C_r}{I_z v_x} w - \frac{l_f C_f + l_r C_r}{I_z} \beta + \frac{l_f C_f}{I_z} \delta_f \\ \dot{\beta} = (-\frac{l_f C_f + l_r C_r}{m v_x^2} - 1) w - \frac{C_f + C_r}{m v_x} \beta + \frac{C_f}{m v_x} \delta_f \end{cases} \tag{7}$$

The control input and output variables of the controlled vehicle are the front wheel steering angle δ_f and the yaw velocity w, respectively. According to the vehicle dynamics model, a relation between the output and the input of the controlled system is deduced

$$\ddot{w}(t) = -a_1 \dot{w}(t) - a_2 w(t) + b_1 \dot{\delta}_f(t) + b_2 \delta_f(t) \tag{8}$$

where $a_1 = \frac{m(l_f^2 C_f + l_r^2 C_r) + I_z(C_f + C_r)}{m v_x I_z}$, $a_2 = \frac{-m v_x^2 (l_f C_f - l_r C_r) + C_f C_r}{m v_x^2 I_z}$, $b_1 = \frac{l_f C_f}{I_z}$, $b_2 = \frac{C_f C_r L}{m v_x I_z}$.

So, the vehicle control system can be described as the following second-order system:

$$\begin{cases} \dot{x}_1 = x_2 \\ \dot{x}_2 = F(x_1, x_2, \dot{\delta}_f, \gamma(t)) + b_0 \delta_f \\ y = x_1 \end{cases} \tag{9}$$

where $x_1 = w(t)$, $x_2 = \dot{w}(t)$, $b_0 = b_2$, $F(x_1, x_2, \dot{\delta}_f, \gamma(t)) = -a_1 \dot{w}(t) - a_2 w(t) + b_1 \dot{\delta}_f(t) + \gamma(t)$, and $F(\cdot)$ is defined as a generalized disturbance that includes the modeled and unmodeled parts of the vehicle and an unknown external disturbance.

3.2. Multi-Dimensional Taylor Network Model

If the function is $\hat{m} + 1$-order differentiable everywhere in the neighborhood at some point, the power series of the eigenfunction expansion at this point is not greater than $\hat{m}$ power in terms of the principle of the multivariate Taylor formula. Based on the MTN model, as shown in Figure 2, the general dynamic equation of an $\hat{t}$-dimensional system can be expressed as [17]:

$$\hat{y}(k+1) = \sum_{\hat{p}=1}^{\hat{N}(\hat{t},\hat{m})} \hat{w}_{\hat{p}}(k) \prod_{\hat{q}=1}^{\hat{t}} \hat{z}_{\hat{q}}^{\hat{\lambda}(\hat{p},\hat{q})}(k) \tag{10}$$

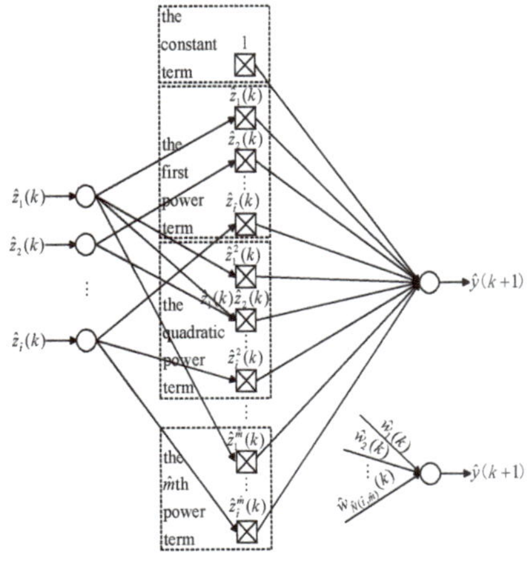

Figure 2. Diagram of the MTN.

In (10), $\hat{N}(\hat{t}, \hat{m})$ represents the total number of product items of the $\hat{t}$-ary function $\hat{f}[\cdot]$, which is expanded into the approximate polynomial with $\hat{m}$ powers; $\hat{w}_{\hat{p}}(k)$ represents the weight coefficient of the $\hat{p}$-th product item in the formula; $\hat{\lambda}(\hat{p},\hat{q})$ represents the power of the variable $\hat{z}_{\hat{q}}(k)$ in the $\hat{p}$-th product item; and $\sum_{\hat{q}=1}^{\hat{t}} \hat{\lambda}(\hat{p},\hat{q}) \leq \hat{m}$, where $\hat{p} = 1,2,\cdots,\hat{N}(\hat{t},\hat{m})$.

As can be seen from Figure 2, the MTN is composed of three main layers, including the input layer, the middle layer, and the output layer. Its structure is simple, it contains only addition and multiplication operations, it is equivalent to only one neuron of the neural network, and it has low algorithm complexity, namely real-time performance. Otherwise, the model parameters are obtained through continuous optimization, which has a "self-learning" ability and good robustness.

The MTN model has a powerful approximation capability for nonlinear systems and can approximate any model with sufficient precision as long as $\hat{N}(\hat{t},\hat{m})$ is large enough [21,39].

4. MTN Predictive Control Scheme with the Differential Term

4.1. Scheme of the MTN Predictive Model

4.1.1. MTN Predictive Model

The MTN model is essentially a one-step-ahead predictive model. Therefore, the MTN model can be regarded as a predictive model. We adopt the MTN model as shown in (10).

4.1.2. Learning Algorithm of the MTN Predictive Model

The BP algorithm is used to adjust the weight coefficients of the MTN model [40,41]. Define the cost function as

$$J_{\mathrm{P}}(k) = \frac{1}{2} e_{\mathrm{P}}{}^{2}(k) \tag{11}$$

where $e_{\mathrm{P}}(k)$ represents the predictive model error and $e_{\mathrm{P}}(k) = y(k) - \hat{y}(k)$; $y(k)$ is the system output; and $\hat{y}(k)$ is the predictive model output.

Let

$$\hat{w}_{\mathrm{P}}(k) = \left(\hat{w}_{\mathrm{P}1}(k), \hat{w}_{\mathrm{P}2}(k), \cdots, \hat{w}_{\mathrm{P}\hat{N}(\hat{t},\hat{m})}(k)\right)^{\mathrm{T}}$$

The BP algorithm can be obtained as follows

$$w_{\mathrm{P}}(k+1) = w_{\mathrm{P}}(k) + \mu \frac{\partial J_P}{\partial w_{\mathrm{P}}(k)} \tag{12}$$

where μ is the learning factor and $\mu > 0$.

4.2. Design of the Predictive Control Law with the Differential Term

Consider the following quadratic performance index:

$$J_{\mathrm{C}}(k) = \min\left\{\frac{1}{2}(r(k) - \hat{y}(k))^2 + \frac{1}{2}\varepsilon \Delta u^2(k)\right\} \tag{13}$$

where $r(k)$ is the reference input, $\hat{y}(k)$ is the output of the MTN predictive model, ε is the control effort weighting factor, $\Delta u(k) = u(k) - u(k-1)$, and $u(k)$ is the control input.

Denote $\frac{\partial J_{\mathrm{C}}(k)}{\partial u(k)} = 0$. We can now obtain the MTN predictive control law:

$$u(k) = u(k-1) + \frac{1}{\varepsilon}(r(k) - \hat{y}(k))\frac{\partial \hat{y}(k)}{\partial u(k)} \tag{14}$$

where $u(k)$, $r(k)$, $\hat{y}(k)$, and ε have the same meaning as in Equation (13).

The differential term is introduced under the traditional quadratic performance index. A differential term can help the system to move in advance, reduce the overshoot, and shorten the adjustment time. Meanwhile, it can reduce the influence of time variation and disturbance factors. The following quadratic performance index with the differential term was adopted:

$$J_C(k) = \min\left\{\frac{1}{2}e_C^2(k) + \frac{1}{2}\varepsilon\Delta u^2(k) + \frac{1}{2}\gamma[e_C(k) - e_C(k-1)]^2\right\} \tag{15}$$

where γ is the differential weighting factor and $\gamma > 0$, $e_C(k) = r(k) - \hat{y}(k)$, $r(k)$, $\hat{y}(k)$, and ε have the same meaning as in Equation (13).

According to (15),

$$\begin{aligned}\frac{\partial J_C(k)}{\partial u(k)} &= e_C(k)\frac{\partial e_C(k)}{\partial u(k)} + \varepsilon\Delta u + \gamma[e_C(k) - e_C(k-1)]\frac{\partial e_C(k)}{\partial u(k)} \\ &= -e_C(k)\frac{\partial \hat{y}(k)}{\partial u(k)} + \varepsilon\Delta u - \gamma[e_C(k) - e_C(k-1)]\frac{\partial \hat{y}(k)}{\partial u(k)}\end{aligned} \tag{16}$$

where ε, γ, $r(k)$, and $\hat{y}(k)$ have the same meaning as in (15).

Denote

$$\frac{\partial J_C(k)}{\partial u(k)} = 0$$

We can now obtain the MTN predictive control law with the differential term:

$$u(k) = u(k-1) + \frac{1}{\varepsilon}e_C(k)\frac{\partial \hat{y}(k)}{\partial u(k)} + \frac{\gamma}{\varepsilon}[e_C(k) - e_C(k-1)]\frac{\partial \hat{y}(k)}{\partial u(k)} \tag{17}$$

where ε, γ, $r(k)$, and $\hat{y}(k)$ have the same meaning as in (15).

4.3. Stability Analysis

Theorem 1. *Consider the quadratic performance index shown in Equation (13). Denote* $h_{\max} = \max_k(h(k))$ *and* $h(k) = \left|\frac{\partial \hat{y}(k)}{\partial u(k)}\right|$. *When* ε *satisfies* $0 < \frac{1}{\varepsilon} < \frac{2}{h_{\max}^2}$, *the closed-loop system is stable.*

Proof. Define $V(k) = \frac{1}{2}e_C^2(k)$. So:

$$\begin{aligned}\Delta V(k) &= V(k+1) - V(k) \\ &= \tfrac{1}{2}[e_C^2(k+1) - e_C^2(k)] \\ &= \tfrac{1}{2}[e_C(k+1) - e_C(k)][e_C(k+1) + e_C(k)] \\ &= \tfrac{1}{2}\Delta e_C(k)[e_C(k+1) + e_C(k)]\end{aligned} \tag{18}$$

$\Delta e_C(k)$ can be denoted by [42,43]

$$\Delta e_C(k) = -\frac{\partial \hat{y}(k)}{\partial u(k)}\Delta u(k) \tag{19}$$

Substitution of (19) into (18) yields

$$\begin{aligned}\Delta V(k) &= \tfrac{1}{2}\left[-\frac{\partial \hat{y}(k)}{\partial u(k)}\Delta u(k)\right][e_C(k+1) + e_C(k)] \\ &= -\tfrac{1}{2}\frac{\partial \hat{y}(k)}{\partial u(k)}\Delta u(k)\left[2e_C(k) - \frac{\partial \hat{y}(k)}{\partial u(k)}\Delta u(k)\right]\end{aligned} \tag{20}$$

According to Equation (14), we can obtain

$$\begin{aligned}\Delta u(k) &= \tfrac{1}{\varepsilon}(r(k) - \hat{y}(k))\frac{\partial \hat{y}(k)}{\partial u(k)} \\ &= \tfrac{1}{\varepsilon}e_C(k)\frac{\partial \hat{y}(k)}{\partial u(k)}\end{aligned} \tag{21}$$

Substituting (21) into (20), we can obtain

$$\begin{aligned}\Delta V(k) &= -\frac{1}{2}\frac{1}{\varepsilon}[e_C(k)]^2\left[\frac{\partial\hat{y}(k)}{\partial u(k)}\right]^2\left\{2-\frac{1}{\varepsilon}\left[\frac{\partial\hat{y}(k)}{\partial u(k)}\right]^2\right\} \\ &\leq -\frac{1}{2}\frac{1}{\varepsilon}[e_C(k)]^2\left[\frac{\partial\hat{y}(k)}{\partial u(k)}\right]^2\left(2-\frac{1}{\varepsilon}\left|\frac{\partial\hat{y}(k)}{\partial u(k)}\right|^2_{\max}\right) \\ &= -\frac{1}{2}\frac{1}{\varepsilon}[e_C(k)]^2\left[\frac{\partial\hat{y}(k)}{\partial u(k)}\right]^2\left(2-\frac{1}{\varepsilon}h^2_{\max}\right)\end{aligned} \tag{22}$$

It can be found that $\Delta V(k) < 0$ if $2-\frac{1}{\varepsilon}h^2_{\max} > 0$ holds. That is, $0 < \frac{1}{\varepsilon} < \frac{2}{h^2_{\max}}$. By the Lyapunov stability theorem, the closed-loop system is stable.

What needs to be noted is that for a quadratic performance index with a differential term, the differential term can help the system to act in advance, reduce the overshoot, and shorten the adjustment time. That is, it can reach the stable region as soon as possible.

It should be noted that an arbitrary number of learning weighting factors was selected according to the stability condition. In an actual situation, the appropriate parameters should be selected according to the actual situation, and the stability condition should be satisfied. □

4.4. Scheme of Predictive Control with the Differential Term Based on the MTN Model

An improved predictive control scheme on the basis of the MTN is proposed for tracking control of SUGVs. First, a predictive model is constructed as follows: (1) the MTN is used as a predictive model to construct a SUGV model; and (2) the BP algorithm is used as its learning algorithm. Second, a predictive control law is designed to control the SUGV. The frame diagram of the improved MTN predictive control system is shown in Figure 3. The front wheel steering angle δ_f is used as the control input, which is equivalent to the $u(k)$ mentioned above. The yaw velocity w is used as the system output, which is equivalent to the $y(k)$ mentioned above. The predictive model output $\hat{w}$ is equivalent to the $\hat{y}(k)$ mentioned above. The desired yaw velocity w_{des} is used as the system input, which is equivalent to the $r(k)$ mentioned above.

The specific steps are as shown in Algorithm 1.

Algorithm 1 Scheme of MTN Predictive Control with the Differential Term

Step 1: Construct MTN predictive model;
(1) Initialize system input and output;
(2) Ascertain structure of MTN predictive model;
(3) Update $w_P(k)$ with Equation (12);
Step 2: Construct MTN predictive control law;
(1) Construct the quadratic performance index, shown as (13);
(2) Update $u(k)$ with Equation (14);
Step 3: Construct an improved predictive control law to control SUGV based on MTN predictive control law;
(1) Construct the quadratic performance index with differential term, shown as (15);
(2) Update the $u(k)$ with Equation (17);
Step 4. Implement the tracking control for the SUGV.

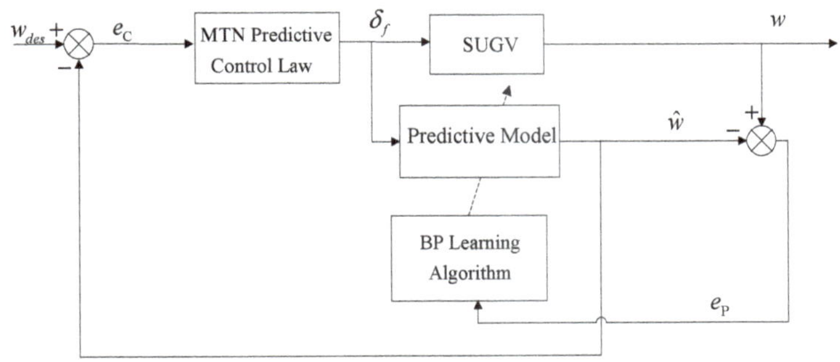

Figure 3. Block diagram of the improved MTN predictive control system.

5. Simulation Results and Discussion

5.1. Simulation Results

SUGVs have a wide range of applications, such as delivery, sanitation, and mining. A tracking control experiment on a SUGV is described here [44,45]. Moreover, the traditional MTN predictive control scheme [24] and the recurrent radial basis (RBF) model predictive control scheme [46] are used for comparison. According to the vehicle dynamics model, as shown in (7), the tracking effect is demonstrated by accurately tracking the desired yaw velocity signal. The control input and output variables are the front wheel steering angle δ_f and the yaw velocity w, respectively. The mathematical relationship between the desired path curve $y = f(x)$ and the desired yaw velocity signal is as follows [44,45]:

$$w_{des} = v_{des} k_{des} \tag{23}$$

$$k_{des} = \frac{d^2y/dx^2}{(1+(dy/dx)^2)^{3/2}} \tag{24}$$

where w_{des} is the desired yaw velocity, v_{des} is the desired speed, and k_{des} represents the curvature of the desired road.

Specific parameters of the SUGV were set as follows: $M = 0.1$ kg, $l_f = 0.1$ m, $l_r = 0.1$ m, $I_z = 0.4\ \mathrm{kg/m^2}$, $v_x = 1$ m/s, $L = 0.6$ m, $C_f = 0.6$ N/rad, and $C_r = 0.6$ N/rad.

Two cases were considered: without noise and with noise. The same parameters were adopted for each of them to verify the performance of the proposed scheme.

Remark 1. *In the following, D-MTNC is our proposed scheme; MTNC is the traditional MTN predictive control scheme; and RBFC is the RBF model predictive control scheme.*

Remark 2. *In the following figures, wdes represents the given reference signal; yMTNC is the traditional MTN predictive control scheme; yD-MTNC is the proposed predictive control scheme; yRBFC is the RBF model predictive control scheme; and eMTNC, eD-MTNC, and eRBFC represent the corresponding tracking errors.*

We consider the control effect of the yaw velocity tracking step signal. The given reference signal is $w_{des} = 1$. The parameters of D-MTNC, MTNC, and RBFC were set as follows.

The MTN predictive model adopts a 4-15-1 structure with four inputs and two powers. The initial value of the weight coefficient vector for the BP algorithm is $\hat{w}_P(0) = 0$ and the learning factor is $\mu = 0.5$.

For the D-MTNC control law parameters, the control effort weighting factor was set to $1/\varepsilon = 0.1$ and the differential term weighting factor was set to $\gamma = 0.005$.

For the MTNC control law parameters, the control effort weighting factor was set to $1/\varepsilon = 0.1$.

The RBF predictive model adopts a 4-18-1 structure. The initial value of the weight coefficient vector for the BP algorithm is $\hat{w}_P(0) = 0$ and the learning factor is $\mu = 0.9$.

For the RBFC control law parameters, the control effort weighting factor was set to $1/\varepsilon = 0.9$.

5.1.1. Without Noise

Figure 4 shows the tracking trajectory and Figure 5 shows the tracking error. Figures 4 and 5 show that the SUGV reaches the desired speed quickly and accurately. Our proposed scheme tracks the given reference signal at $k = 5$, whereas the traditional MTN predictive control scheme and the RBF predictive control scheme track the reference signal at $k = 70$ and $k = 250$, respectively. These results verify that the improved predictive control scheme has a better control effect and faster convergence than the other schemes.

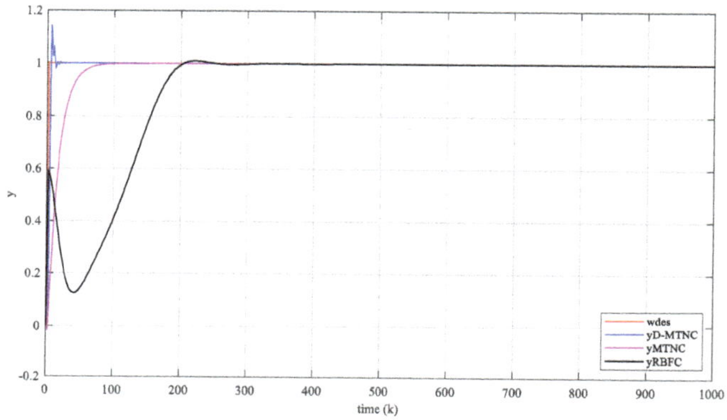

Figure 4. The tracking trajectory without noise.

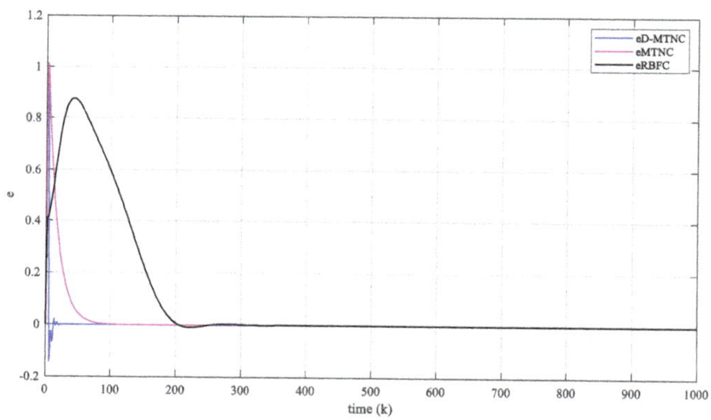

Figure 5. The tracking error without noise.

5.1.2. With Noise

To verify the robustness of the proposed control scheme, Gaussian white noise and 100 times the noise were added to the system. Regarding the Gaussian white noise, it exhibits a mean of 0 and a standard deviation of 0.2 at the 300th time point.

Figure 6 shows the tracking trajectory and Figure 7 shows the tracking error. From Figures 6 and 7, when 100 times the Gaussian white noise is added at the 300th time point to the system, there is a sharp change at the 300th time point, but the proposed control scheme has a quick response, which retracks the reference more quickly than the other schemes. These results verify that the improved predictive control scheme has a good control effect and good robustness.

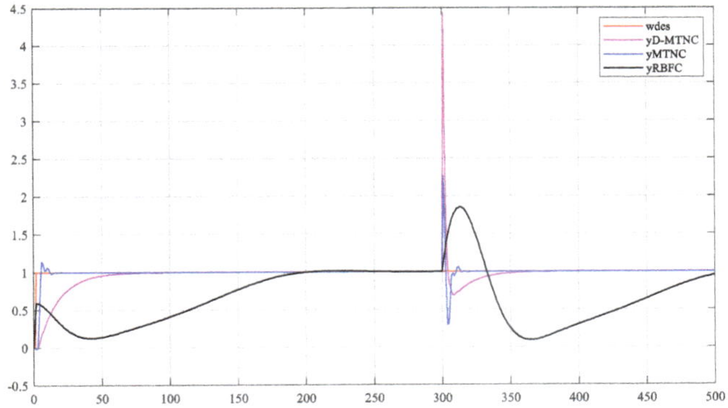

Figure 6. The tracking trajectory with noise.

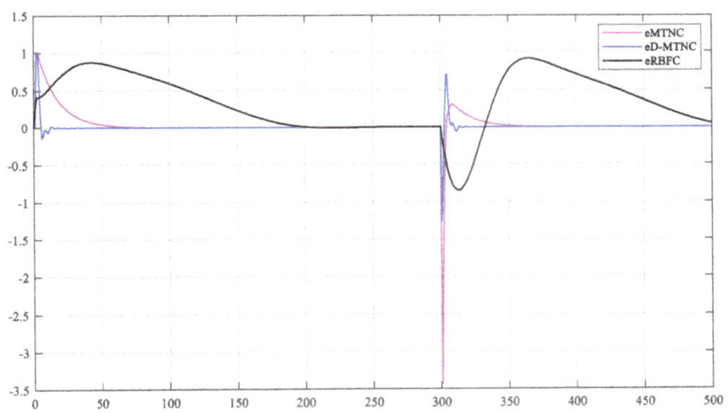

Figure 7. The tracking error with noise.

5.2. Discussion

For tracking control of SUGVs, the above results show that the proposed method has good performance and tracks the reference signal quickly and accurately. To verify the robustness of the proposed scheme, sharp noise was added. The results of the experiment in terms of tracking, robustness, and real-time performance all show that the proposed scheme has better performance than MTNC and RBFC.

Regarding the choice of model structure, what needs illustration is that the MTN predictive model adopted a 4-15-1 structure and the RBF predictive model adopted a 4-18-1 structure in this paper. The MTN structure depends on the middle layer's nodes when the input and output layer are confirmed. The middle layer's nodes rely on the input layer's nodes and the power item. The MTN predictive model's structure is shown in Figure 1. From Figure 1, when the input layer had four nodes and the number of power items was 2, the number of product items was 15. The RBF hidden layer's nodes depend on the history of multiple experiments, and we chose 18 nodes for the hidden layer here.

The proposed scheme showed good performance in tracking control of the SUGV. However, SUGVs have a wide range of applications, such as delivery, sanitation, and mining. When facing these practical scenarios, the proposed method needs to be experimentally validated and the appropriate parameters need to be selected according to the actual situation.

6. Conclusions

An improved predictive control scheme based on the MTN was proposed for tracking control of SUGVs. The traditional objective function was improved to obtain a predictive objective function with the differential term. The optimal control quantity was given in real time through iterative optimization. A tracking control experiment on a SUGV was carried out to verify the effectiveness of the proposed scheme. The results show that the proposed scheme is effective and has good real-time, robustness, and convergence performance, which ensure that the vehicle can quickly and accurately track the desired yaw velocity signal, and is superior to the traditional MTN and RBF predictive control schemes.

The proposed improved MTN predictive control scheme developed in the present paper shows great promise but requires further study. For example, the proposed scheme could be applied to different complex driving scenarios, and some complicated conditions could be considered, such as an actuator failure or input constraints.

Author Contributions: Writing—original draft preparation, Y.W.; methodology, C.L.; writing—review and editing, C.Y.; validation, M.L.; formal analysis, H.L. All authors have read and agreed to the published version of the manuscript.

Funding: This research received no external funding.

Institutional Review Board Statement: Not applicable.

Informed Consent Statement: Not applicable.

Data Availability Statement: Not applicable.

Acknowledgments: The authors thank the Editor-in-Chief, the Associate Editor, and the anonymous reviewers for their valuable comments and suggestions.

Conflicts of Interest: The authors declare no conflict of interest.

References

1. Wang, H.B.; Zuo, J.S.; Liu, S.D.; Zheng, W.; Wang, L. Model-free adaptive control of steady-state drift of unmanned vehicles. *Control Theory Appl.* **2021**, *38*, 23–32. (In Chinese)
2. Elbanhawi, M.; Simic, M.; Jazar, R. In the passenger seat: Investigating ride comfort measures in autonomous cars. *IEEE Intel. Transp. Syst. Mag.* **2015**, *7*, 4–17. [CrossRef]
3. Kim, S.; Gwon, G.P.; Hur, W.S.; Hyeon, D.; Kim, D.-Y.; Kim, S.-H.; Kim, D.-K.; Lee, S.-H.; Lee, S.; Shin, M.-O.; et al. Autonomous campus mobility services using driverless taxi. *IEEE Trans. Intel. Transp. Syst.* **2017**, *18*, 3513–3526. [CrossRef]
4. Kramberger, T.; Potocnik, B. LSUN-stanford car dataset: Enhancing large-scale car image datasets using deep learning for usage in GAN training. *Appl. Sci.* **2020**, *10*, 4913. [CrossRef]
5. Shen, K.; Wang, M.; Fu, M.; Yang, Y.; Yin, Z. Observability analysis and adaptive information fusion for integrated navigation of unmanned ground vehicles. *IEEE Trans. Ind. Electron.* **2020**, *67*, 7659–7668. [CrossRef]
6. Wang, H.; Yuan, S.; Guo, M.; Chan, C.-Y.; Li, X.; Lan, W. Tactical driving decisions of unmanned ground vehicles in complex highway environments: A deep reinforcement learning approach. *Proc. Inst. Mech. Eng. Part D J. Automob. Eng.* **2021**, *235*, 1113–1127. [CrossRef]
7. Zhang, H.; Zhang, H.M.; Wang, Z.P. Event-triggered predictive path following control for unmanned autonomous vehicle. *Control Decis.* **2019**, *34*, 2421–2427. (In Chinese)
8. Kaur, K.; Rampersad, G. Trust in driverless cars: Investigating key factors influencing the adoption of driverless cars. *J. Eng. Technol. Manag.* **2018**, *48*, 87–96. [CrossRef]
9. Chen, G.; Zhang, W.G.; Zhang, X.N. Speed tracking control of a vehicle robot driver system using multiple sliding surface control schemes. *Int. J. Adv. Robot Syst.* **2013**, *10*, 1–9. [CrossRef]
10. Song, C.; Zhao, R.; Yang, H.; Zeng, Z.; Wu, J. Using fuzzy control for parallel-inverter system with nonlinear-load. In Proceedings of the 2010 International Conference on Electrical Machines and Systems, Incheon, Korea, 10–13 October 2010; pp. 193–197.
11. Niu, H.; Bhowmick, C.; Jagannathan, S. Attack detection and approximation in nonlinear networked control systems using neural networks. *IEEE Trans. Neural Netw. Learn. Syst.* **2020**, *31*, 235–245. [CrossRef]
12. Zhao, J.; Sun, J.; Cai, Z.; Wang, L.; Wang, Y. End-to-End deep reinforcement learning for image-based UAV autonomous control. *Appl. Sci.* **2021**, *11*, 8419. [CrossRef]
13. Mehmood, A.; Zameer, A.; Ling, S.H.; Rehman, A.U.; Raja, M.A.Z. Integrated computational intelligent paradigm for nonlinear electric circuit models using neural networks, genetic algorithms and sequential quadratic programming. *Neural Comput. Appl.* **2020**, *32*, 10337–10357. [CrossRef]

14. Wu, Z.; Rincon, D.; Christofides, P.D. Process structure-based recurrent neural network modeling for model predictive control of nonlinear processes. *J. Process Control* **2020**, *89*, 74–84. [CrossRef]
15. Piche, S.; Sayyar-Rodsari, B.; Johnson, D.; Gerules, M. Nonlinear model predictive control using neural networks. *IEEE Control Syst. Mag.* **2000**, *20*, 53–62.
16. Rummelhart, D.E.; Hinton, G.E.; Williams, R.J. Learning representations by back propagating errors. *Nature* **1986**, *323*, 533–536. [CrossRef]
17. Zhou, B.; Yan, H.S. Financial time series forecasting based on wavelet and multi-dimensional Taylor network dynamics model. *Syst. Eng.-Theory Pract.* **2013**, *33*, 2654–2662. (In Chinese)
18. Yan, H.S.; Kang, A.M. Asymptotic tracking and dynamic regulation of SISO nonlinear system based on discrete multi-dimensional Taylor network. *IET Control Theory Appl.* **2017**, *11*, 1619–1626. [CrossRef]
19. Lin, Y.; Yan, H.S.; Zhou, B. Nonlinear time series prediction method based on multi-dimensional Taylor network and its applications. *Control Decis.* **2014**, *29*, 795–801. (In Chinese)
20. Li, C.L.; Yan, H.S. Nonlinear time-delay system identification and prediction based on multi-dimensional Taylor network and IPSO. *J. Grey Syst.* **2018**, *30*, 96–109.
21. Li, C.L.; Yan, H.S.; Zhang, J.J. Multi-dimensional Taylor network adaptive predictive control for singleinput single-output nonlinear systems with input time-delay. *Trans. Inst. Meas. Control* **2022**, *44*, 595–608. [CrossRef]
22. Zhang, J.J.; Yan, H.S. MTN optimal control of MIMO non-affine nonlinear time-varying discrete systems for tracking only by output feedback. *J. Frankl. Inst.* **2019**, *356*, 4304–4334. [CrossRef]
23. Zhang, C.; Yan, H.S. Identification and adaptive Multi-dimensional Taylor network control of single-input single-output non-linear uncertain time-varying systems with noise disturbances. *IET Control Theory Appl.* **2019**, *13*, 841–853. [CrossRef]
24. Li, C.L.; Ma, X.S.; Zhang, J.J. Nonlinear system predictive control using a multi-dimensional Taylor network. *Trans. Inst. Meas. Control* **2019**, *41*, 3396–3405. [CrossRef]
25. Schwarting, W.; Alonso-Mora, J.; Paull, L.; Karaman, S.; Rus, D. Safe nonlinear trajectory generation for parallel autonomy with a dynamic vehicle model. *IEEE Trans. Intel. Transp.* **2018**, *19*, 2994–3008. [CrossRef]
26. Li, R.; Zhang, L.; Han, L.; Wang, J. Multiple vehicle formation control based on robust adaptive control algorithm. *IEEE Intel. Transp. Syst.* **2017**, *9*, 41–51. [CrossRef]
27. Zhen, Z.; Tao, G.; Yu, C.; Xue, Y. A multivariable adaptive control scheme for automatic carrier landing of UAV. *Aerosp. Sci. Technol.* **2019**, *92*, 714–721. [CrossRef]
28. Tagne, G.; Talj, R.; Charara, A. Higher-order sliding mode control for lateral dynamics of autonomous vehicles, with experimental validation. In Proceedings of the 2013 IEEE Intelligent Vehicles Symposium (IV), Gold Coast, QLD, Australia, 23–26 June 2013; pp. 678–683.
29. Wang, P.; Gao, S.; Li, L.; Cheng, S.; Zhao, L. Automatic steering control strategy for unmanned vehicles based on robust backstepping sliding mode control theory. *IEEE Access* **2019**, *7*, 64984–64992. [CrossRef]
30. Yu, L.; Yan, X.; Kuang, Z.; Chen, B.; Zhao, Y. Driverless bus path tracking based on fuzzy pure pursuit control with a front axle reference. *Appl. Sci.* **2020**, *10*, 230. [CrossRef]
31. Chen, Y.; Hu, C.; Qin, Y.; Li, M.; Song, X. Path planning and robust fuzzy output-feedback control for unmanned ground vehicles with obstacle avoidance. *Proc. Inst. Mech. Eng. Part D J. Automob. Eng.* **2021**, *235*, 933–944. [CrossRef]
32. Ji, X.W.; He, X.K.; Chen, L.; Liu, Y.; Wu, J. Adaptive-neural-network-based robust lateral motion control for autonomous vehicle at driving limits. *Control Eng. Pract.* **2018**, *76*, 41–53. [CrossRef]
33. Kabzan, J.; Hewing, L.; Liniger, A.; Zeilinger, M.N. Learning-based model predictive control for autonomous racing. *IEEE Robot Autom. Lett.* **2019**, *4*, 3363–3370. [CrossRef]
34. Zhao, K.; Wang, C.; Xaio, G.; Li, H.; Ye, J.; Liu, Y. Research for nonlinear model predictive controls to laterally control unmanned vehicle trajectory tracking. *Appl. Sci.* **2020**, *10*, 6034. [CrossRef]
35. Kayacan, E.; Ramon, H.; Saeys, W. Robust trajectory tracking error model-based predictive control for unmanned ground vehicles. *IEEE-ASME Trans. Mech.* **2016**, *21*, 1–9. [CrossRef]
36. Bozek, P.; Karavaev, Y.L.; Ardentov, A.A. Neural network control of a wheeled mobile robot based on optimal trajectories. *Int. J. Adv. Robot Syst.* **2020**, *17*, 1–10. [CrossRef]
37. Cummings, M.L.; How, J.P.; Whitten, A.; Toupet, O. The impact of human-automation collaboration in decentralized multiple unmanned vehicle control. *Proc. IEEE* **2012**, *100*, 660–671. [CrossRef]
38. Donmez, B.; Nehme, C.; Cummings, M.L. Modeling workload impact in multiple unmanned vehicle supervisory control. *IEEE Trans. Syst. Man Cybern.-Part A Syst. Hum.* **2010**, *40*, 1180–1190. [CrossRef]
39. Klambauer, G. *Mathematical Analysis*; Marcel Dekker Inc.: New York, NY, USA, 1975.
40. Cowan, C.F.N.; Grant, P.M. (Eds.) *Adaptive Filters*; Prentice-Hall, Inc.: Englewood Cliffs, NJ, USA, 1985.
41. Al-Batah, M.S.; Isa, N.A.M.; Zamli, K.Z.; Azizli, K.A. Modified recursive least squares algorithm to train the hybrid multilayered perceptron (HMLP) network. *Appl. Soft Comput.* **2010**, *10*, 236–244. [CrossRef]
42. Ku, C.C.; Lee, K.Y. Diagonal recurrent neural networks for dynamic systems control. *IEEE Trans. Neural Netw.* **1995**, *6*, 144–156.
43. Peng, J.Z.; Dubay, R. Identification and adaptive neural network control of a DC motor system with dead-zone characteristics. *ISA Trans.* **2011**, *50*, 588–598. [CrossRef] [PubMed]

44. Nguyen, T.W.; Catoire, L.; Garone, E. Control of a quadrotor and a ground vehicle manipulating an object. *Automatica* **2019**, *105*, 384–390. [CrossRef]
45. Ren, H.; Karimi, H.R.; Lu, R.; Wu, Y. Synchronization of network systems via aperiodic sampled-data control with constant delay and application to unmanned ground vehicles. *IEEE Trans. Ind. Electron.* **2020**, *67*, 4980–4990. [CrossRef]
46. Tavoosi, J.; Mohammadzadeh, A. A new recurrent radial basis function network-based model predictive control for a power plant Boiler. *Int. J. Eng.* **2021**, *34*, 667–675.

applied sciences

Article

Design of a NARX-ANN-Based SP Controller for Control of an Irrigation Main Canal Pool

Ybrain Hernandez-Lopez [1], Raul Rivas-Perez [1,*] and Vicente Feliu-Batlle [2,*]

1 Departamento de Automática y Computación, Universidad Tecnológica de la Habana José Antonio Echeverría (CUJAE), La Habana 19390, Cuba
2 Escuela Técnica Superior de Ingeniería Industrial de Ciudad Real, Universidad de Castilla-La Mancha (UCLM), Av. Camilo José Cela s/n, 13071 Ciudad Real, Spain
* Correspondence: raul_rivas_perez@yahoo.es (R.R.-P.); vicente.feliu@uclm.es (V.F.-B.)

Abstract: The management of irrigation main canals are studied in this research. One way of improving this is designing an efficient automatic control system of the water that flows through the canal pools, which is usually carried out by PI controllers. However, since canal pools are systems with large time delays and nonlinear hydrodynamics, these PIs are tuned in a very conservative way so that the closed-loop instability that may appear depending on the chosen operation regime is avoided. These controllers are inefficient because they have slow time responses. In order to obtain faster responses that remain stable independently of the operation regime, a control system that combines a Smith predictor, which is appropriate to control linear systems with large time delays, with a NARX artificial neural network (ANN), that models the nonlinear dynamics of the pools, is proposed. By applying system identification procedures, two nonlinear NARX-ANN-based models and a linear mathematical model of a real canal pool were obtained. These models were applied to implement a modified NARX-ANN-based SP controller and a conventional linear SP controller. Experimental results on our real canal pool showed that our modified NARX-ANN-based SP controller overcomes conventional linear SP controllers in both setpoint tracking and load disturbance rejection.

Keywords: NARX-ANN-based models; modified SP controller design; irrigation main canal pool automation; system identification; management of water resources

Citation: Hernandez-Lopez, Y.; Rivas-Perez, R.; Feliu-Batlle, V. Design of a NARX-ANN-Based SP Controller for Control of an Irrigation Main Canal Pool. *Appl. Sci.* **2022**, *12*, 9180. https://doi.org/10.3390/app12189180

Academic Editor: Emanuele Carpanzano

Received: 28 July 2022
Accepted: 8 September 2022
Published: 13 September 2022

1. Introduction

Water is a vital and essential resource for human life, environment, as well as for sustainable development [1]. However, millions of people currently live in conditions of extreme water scarcity [2]. Water scarcity is the consequence of two particularities: physical scarcity, and economic scarcity [3,4]. Physical scarcity is determined as the insufficiency of natural resources to supply a region's demand for water, while economic scarcity is demarcated as the consequence of poor investment in science, technologies, and infrastructures [5,6].

In irrigation systems, most of the water resources are transported by the networks of irrigation canals [7–9]. In most countries, irrigation canals are manually operated [10–12].

Currently, one of the reliable solutions to face the problems related to water scarcity in irrigated agriculture is the automatic control, which is considered a powerful tool to improve the operation and efficiency of water distribution in the irrigation main canals [13,14].

The application of conventional PID (proportional integral derivative) controllers is one of the most widespread solutions to control the water distribution in irrigation main canal pools, see e.g., [15–17]. However, this control strategy usually does not yield satisfactory results because irrigation main canal pools are complex systems that are characterized by nonlinear dynamic behaviors with dominant time delays [18–20], and PID controllers are not adequate for this class of processes [21–24]. Therefore, the design and implementation of controllers aimed at increasing operability and safety in irrigation main canal pools, represents an important and challenging research field [25–27].

One of the most widely used controllers for processes with dominant time delays (processes in which the time delay is much greater than the largest of the time constants) is the Smith predictor (SP) [28–30]. It is characterized by having a linear mathematical model of the process to be controlled in its structure, and it shows a low effectiveness in rejecting load disturbances [31,32].

The SP has been the subject of intensive research and modifications aimed at improving its performance, robustness, and disturbance rejection features, as well as enabling its application in diverse processes (unstable, time-varying, nonlinear, etc.), see e.g., [33–35].

The SP has been applied to water distribution control in different irrigation main canal pools. Among the first proposals are those made by M.J. Shand [36], R. Rivas-Perez [9], J.L. Del-tour [37] and F. Sanfilippo [38]. They are characterized by combining the conventional structure of the SP with PI or PID controllers, which do not guarantee an effective water distribution control. The adaptive SP was proposed also to improve the effectiveness of water distribution control in irrigation main canal pools with large variations in their operating conditions [18,26]. Furthermore, in recent years, the SP has been combined with fractional order controllers to increase the robustness of the control of irrigation canal pools, see e.g., [19,34].

One of the essential problems that present these SP-based control systems is their poor performance due to the use (in the SP structure) of linear mathematical models of the water distribution dynamic behavior in irrigation main canal pools [6]. They also have a low rejection of load disturbances, which originates insufficient accuracy in water distribution control.

Irrigation main canals are nonlinear complex processes [8,9]. The physical dynamics of the water distribution in irrigation main canal pools has traditionally been modeled by the Saint-Venant equations, nonlinear hyperbolic partial differential equations with complex constraints, whose analytical solution is very complicated in the case of an arbitrary geometry [39,40]. These equations are derived from mass and momentum balances and are very difficult to use for the design of a controller [8]. Consequently, it is used in the application of different linearization methods [20] or the obtention of linear mathematical models by using systems identification tools [23,40]. The linear mathematical models yielded by these procedures do not characterize accurately the nonlinear dynamic behavior of the water distribution in irrigation main canal pools and, accordingly, their application to the controller's design does not yield satisfactory results in many cases [6,18].

In general, nonlinear processes can only be adequately characterized by nonlinear models [41]. Thus, in spite of the abundant research in this field, the development of nonlinear control-oriented models of irrigation main canal pools remains an open problem.

Recently, artificial neural networks (ANNs) are being used to describe effectively and accurately the dynamic behavior of processes with nonlinear dynamic behaviors. They are computational models that process information inspired by the behavior observed in biological neurons [42–44].

ANNs receive increasing acceptance and are applied in diverse fields because they use non-physics-based algorithms, handle appropriately nonlinear relationships, and incorporate user experience into their structure [45–47]. Therefore, they constitute a powerful tool for obtaining adequate dynamic models of nonlinear and time-varying processes that can be applied to the design of controllers based on nonlinear process models, including SP and model predictive controllers (MPC), see e.g., [48–52].

An important and useful class of discrete-time nonlinear models is the Nonlinear Autoregressive with eXogenous Inputs (NARX) model, which was introduced by S.A. Billings in 1980s as a new representation of a wide class of discrete nonlinear systems [41]. Different ANN architectures have been considered as subsets of the NARX model [43].

The NARX-ANN and TD-NARX-ANN (Time Delay-NARX-ANN) are recurrent dynamic neural network architectures used to model different nonlinear dynamic systems that are characterized by a good learning, fast convergence, and well generalization [53,54].

Diverse authors have proposed the use of NARX-ANN and TD-NARX-ANN for modelling different nonlinear processes and implementing their control systems. A NARX-ANN model for the prediction of the daily direct solar radiation was used in [55]. A NARX-ANN based modeling approach to establish a robust water management tool to aid urban water managers in controlling the development of shallow water tables induced by artificial recharge activity was presented in [56]. A NARX-ANN model for accurate discharge prediction through lateral orifices of irrigation main canals was employed in [57]. A NARX-ANN to predict moisture sorption kinetics and consequently equilibrium moisture content of shiitake mushrooms over a wide range of relative humidity and temperature was developed in [58]. A NARX model reference adaptive control system of a magnetic levitation process was introduced in [59]. A NARX-based model predictive controller for grain dryers was designed in [60]. A NARX-based model predictive controller for turboshaft direct fuel control of an aircraft engine to improve its response ability in the presence of degraded conditions was implemented in [61]. A novel control system for a shunt active power filter based on a NARX-ANN was proposed in [62]. Based on a NARX model, two intelligent controllers to guarantee the desired growth conditions of crops, such as humidity and temperature, in a greenhouse were implemented in [63]. A general-purpose genetic algorithm (GA)-optimized NARX-ANN-based controller for control of nonlinear chemical and biochemical processes was introduced in [64]. A NARX-ANN-based predictive controller of personalized heating systems was designed in [65]. A NARX-ANN-based Smith predictor controller for time-delay compensation of nonlinear processes was developed in [46,66].

The main objective of this paper is the design of modified SP controllers that use NARX-ANN models as nonlinear internal models for the accurate and reliable control of water distribution in an irrigation main canal pool.

The major contributions of this paper are: (1) the proposal for the first time of a water distribution controller in an irrigation main canal pool founded on a modified SP with a NARX-ANN based model that describes the nonlinear dynamic behavior of the canal pool with an arbitrarily small error and (2) the improvements in the SP rejection features to load disturbances by means of including a gain tuning block $K(t)$ that allows an adequate estimation of the load disturbances to be removed.

The proposed controller has the advantages of increasing the operability and reliability of the water distribution control systems, as well as reducing current water losses due to operation in the irrigation main canals pools.

This article is organized as follows. Section 2 proposes mathematical models of the studied process through the application of systems identification procedures. Section 3 develops the design of the NARX-ANN-based SP controller. Section 4 shows the discussions of the obtained results. Finally, some conclusions are offered in Section 5.

2. ANN-Based System Identification of an Irrigation Main Canal Pool

The research presented in this paper is based on the Zaza-Ciego de Avila Main Canal (ZCAMC), which is one of the most important hydraulic constructions in Cuba [67]. This canal is in the central zone of the country and connects the Zaza reservoir (with a capacity of 1,020,106 m^3) with the south of the Ciego de Avila County. The surface irrigated is about 148,200 ha that is used for the cultivation of rice, sugar cane, citrus fruits, soybeans, horticulture, as well as for the development of fishing and livestock. The canal delivers a total volume of 900,106 m^3 of water/year that alleviates the overexploitation of the phreatic level in the region [68].

This canal is a 44.5 km long cross-structure with variable water discharge between 33.0 and 23.0 m^3/s, overall canal height from 4.80 to 3.87 m, and normal water depth varying from 4.30 to 3.10 m. It has a trapezoidal cross-section, the bottom width varies from 6 to

3.2 m, and the average slope is 0.07 m/km [6]. It has four pools separated by undershot gated cross-structures.

This canal is mostly operated in a downstream end regulation mode [8,26]. All the pools of this canal are electrified and equipped with downstream end water level and gate position sensors, SIMATIC S7-300 PLCs, motors for the gates' positioning, and systems for data acquisition, processing, and storage.

Likewise, this canal has a remote supervisory control and data acquisition system (SCADA) with communication by radio and field buses that provide real time data of the controlled variables. Figure 1 is a view of ZCAMC.

Figure 1. A view of Zaza-Ciego de Avila Main Canal (ZCAMC).

Mathematical models have a vital role in systems analysis and design [41]. The controller design habitually requires a dynamic mathematical model that accurately characterizes the behavior of the processes to be controlled [69,70]. Therefore, one of the significant challenges of the controller design methodology is to obtain a suitable mathematical model of the process.

Several mathematical models have been developed for the irrigation main canal pools, see e.g., [8,40,50]. Since the irrigation main canal pools have complex nonlinear dynamics, linear approximation models of their dynamic processes are often developed [8,9], which have been often attained by applying system identification procedures, see e.g., [9,18,23].

System identification is the approach used to infer and construct process models from a set of experimental input-output data [41,69].

Since 1986, neural networks have been applied in the system identification of nonlinear dynamic processes [42].

Data and results reported in this paper were obtained from the first pool in the ZCAMC. It is a trapezoidal cross section main canal pool 12.5 km long, operated by using the distant downstream operation method [6,70]. This pool is equipped with an industrial personal computer, and a SIMATIC S7-300 PLC. The downstream end water level is controlled by four upstream undershoot gates situated at the beginning of the pool. Figure 2 is a schematic representation of this canal pool, in which the four control gates are represented by an equivalent gate.

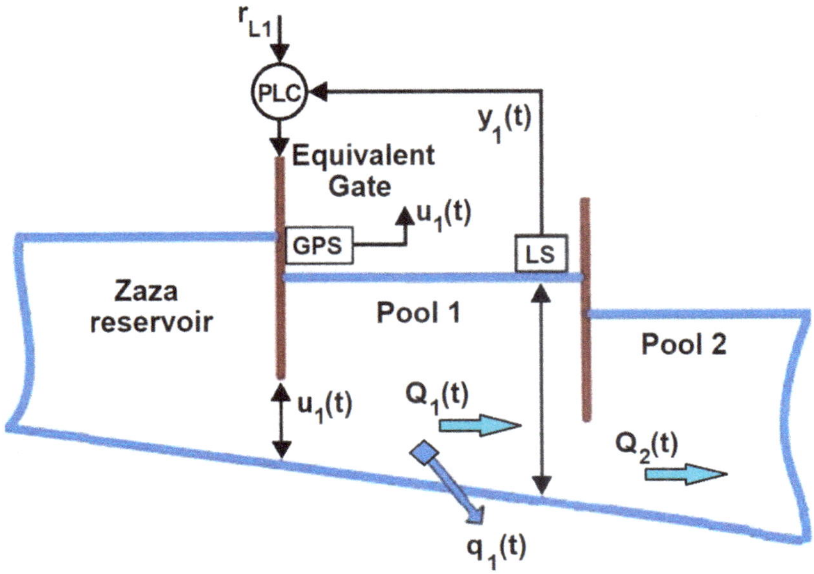

Figure 2. Schematic diagram of the first pool in ZCAMC, where LS is a level sensor, GPS is a gate position sensor, PLC is a programmable logic controller, $Q_i(t)$ are the water flows; $u_1(t)$ is the gate position, $q_1(t)$ is the lateral discharge, r_{L1} is the water level setpoint.

The available measurements are the downstream end water levels $y_1(t)$, and the gate positions $u_1(t)$. Thus, the controlled variable is the downstream end water level $y_1(t)$, and the manipulated variable is the position of the upstream equivalent gate $u_1(t)$.

In this paper, using the system identification procedures [69] and the real-time experimental data attained through the SIMATIC S7-300 PLC system in open loop operating points, three models are obtained that describe the nominal dynamic behavior of the water distribution in the studied canal pool. The first and second models are nonlinear and are based on a NARX-ANN and a TD-NARX-ANN with recurrent architectures [53] due to the advantages of this type of structures, among which are the possibility of approximating any function with the desired precision, provided that it presents a limited number of discontinuities [43].

For nonlinear system identification the Matlab Neural Network-based System Identification Toolbox was used.

Recurrent ANNs are characterized by having feedback (or connections) between neurons in one layer and neurons in previous layers and in the same layer, in addition to forward connections, as well as maintaining a certain dependence on events that occurred in past moments (memory effect), which is convenient in the system identification procedure [43].

The development of the best architecture for the NARX-ANN-based models requires the determination of the number of hidden layers, the activation functions, the process time delays, and the efficient training algorithm [53].

The mathematical formulation of the NARX-ANN-based model can be written by the following discrete time expression [71]:

$$y(t) = f(u(t-1), u(t-2), \ldots, u(t-n_u), y(t-1), y(t-2), \ldots, y(t-n_y)), \quad (1)$$

where $u(t)$ and $y(t)$ denote respectively the input and output of the model at the discrete time step t, n_u, n_y are the input-memory and output-memory orders respectively, and $f(\cdot)$ is a nonlinear function represented by the ANN. The sampling period of the input and output data was T =30 s.

The TD-NARX-ANN-based model is expressed as [72,73]:

$$y(t) = f(u(t-d-1), u(t-d-2), \ldots, u(t-d-n_u), y(t-1), y(t-2), \ldots, y(t-n_y)), \quad (2)$$

where d ($d > 0$) is the number of process time delay sampling periods. We consider $d = 10$ which implies a time delay of 300 s.

Figures 3 and 4 illustrate the diagrams with the recurrent architectures of the NARX-ANN and TD-NARX-ANN-based models that have been designed, which have an input layer with 13 memory blocks for the input and output signals, a hidden layer with three neurons, and an output layer with one neuron. The activation functions in the hidden and output layers are hyperbolic tangent sigmoid and piecewise-linear (symmetric saturating linear) respectively.

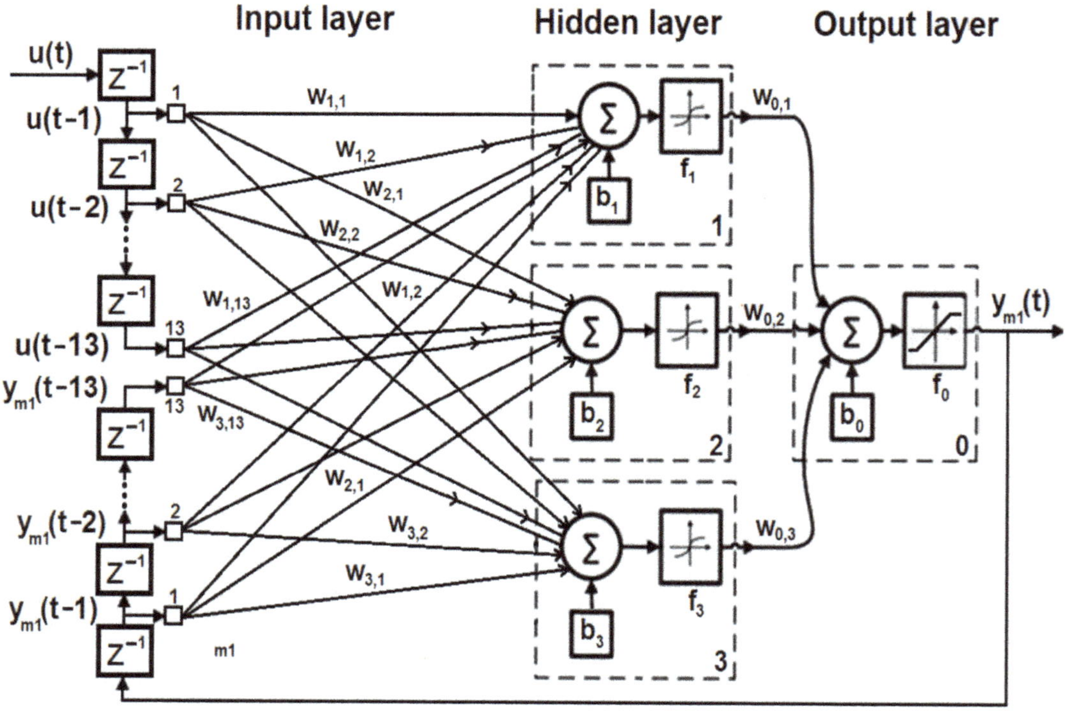

Figure 3. Architecture of the NARX-ANN-based model designed.

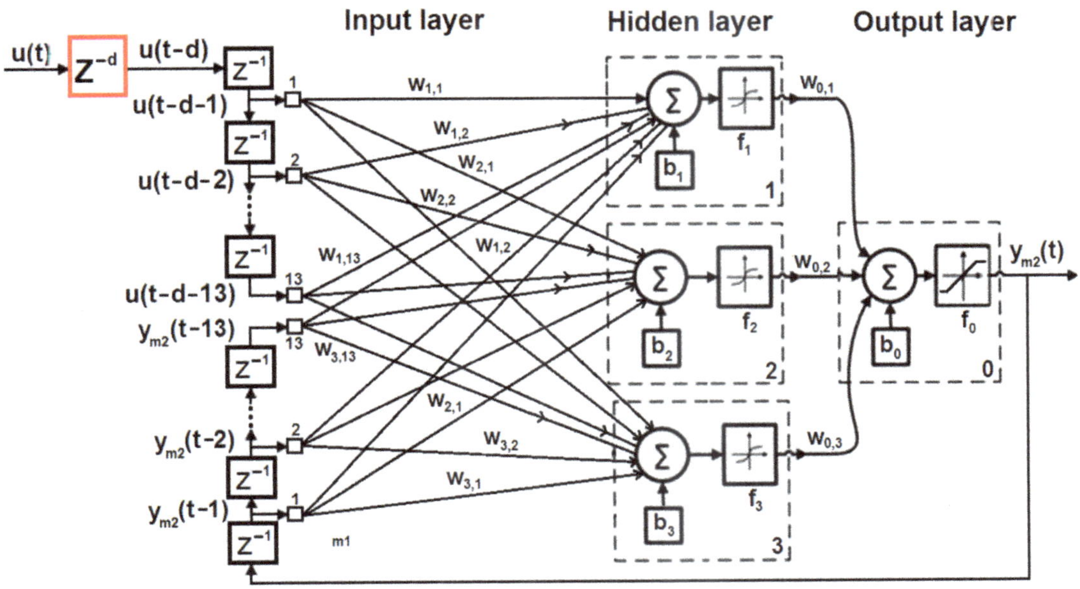

Figure 4. Architecture of the TD-NARX-ANN-based model designed.

The difference between the architecture of these models lies in the TD-NARX-ANN-based model has the time delay block Z^{-d} highlighted in red line.

The mathematical expressions of the NARX-ANN- and TD-NARX-ANN-based models are, respectively, [43]:

$$y_{m1}(t) = f_0\left(\sum_{h=1}^{n_h} w_{0,h} f_h\left(\sum_{i=1}^{n_u} w_{h,i} u(t-i) + \sum_{j=1}^{n_y} w_{h,j} y_{m1}(t-j) + b_h\right) + b_0\right), \quad (3)$$

$$y_{m2}(t) = f_0\left(\sum_{h=1}^{n_h} w_{0,h} f_h\left(\sum_{i=1}^{n_u} w_{h,i} u(t-d-i) + \sum_{j=1}^{n_y} w_{h,j} y_{m2}(t-j) + b_h\right) + b_0\right), \quad (4)$$

where $w_{h,i}$ are the synaptic weights of each $h = 1, \ldots, 3$ neuron of the hidden layer from the $i = 1, 2, \ldots, n_u = 13$ inputs $u(t)$, $w_{h,j}$ are the synaptic weights of each h neuron of the hidden layer from the $j = 1, 2, \ldots, n_y = 13$ outputs $(y(t))$, $w_{0,h}$ are the synaptic weights of the output layer neuron from each h neurons of the hidden layer, b_h are the bias weights of each h neuron of the hidden layer, b_0 is the bias weight of the output layer neuron, f_h is the hyperbolic tangent activation function of each h neuron of the hidden layer, which is defined by the expression:

$$f_h(x) = \frac{e^x - e^{-x}}{e^x + e^{-x}} \quad (5)$$

n_h is the neuron number of the hidden layer, and f_0 is a piecewise-linear activation function of the output layer, which is formed by three linear pieces joined to form a non-linear function (symmetric saturating linear) defined as:

$$f_0(x) = \begin{cases} -1, x < -1, \\ x, -1 < x < 1, \\ 1, x > 1. \end{cases} \quad (6)$$

The optimal number of neurons in the hidden layer was attained by using the Akaike Information Criterion (AIC). The lowest value of AIC $= -6.29$ was obtained for $h = 3$.

The input-memory and output-memory orders ($n_u = n_y = 13$) were obtained from the time-series modeling using the Box–Jenkins approach.

The data set Z^N that describes the irrigation canal pool studied in its entire range of operation was attained by exciting the upstream equivalent undershoot gate, i.e., the process input $u_1(t)$, with a pseudo random binary sequence (PRBS), and measuring the downstream end water level $y_1(t)$:

$$Z^N = \{[u_1(t), y_1(t)] | t = 1, \ldots, N\} \quad (7)$$

Figure 5 illustrates the dataset Z^N obtained, which was split into three parts, 50% for the training set, 25% for the validation set, and 25% for the testing set.

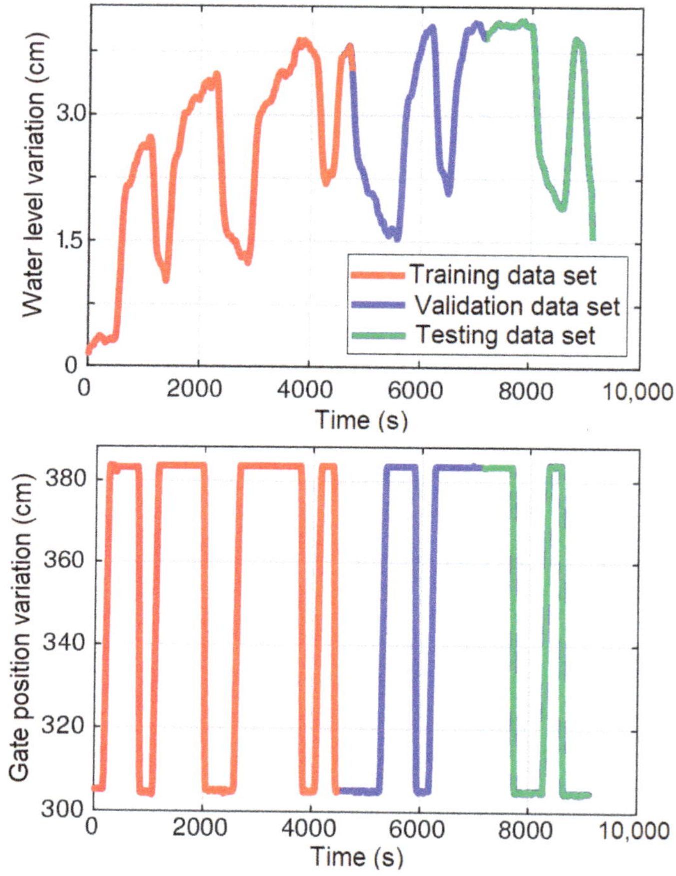

Figure 5. The obtained data set Z^N of the irrigation canal pool studied.

Training is an iterative process that consists of determining the synaptic weights w_h and bias b_h of each of the h neurons of the ANNs designed, so that their output signals converge towards the desired states [72]. The training algorithm should adjust the ANN parameters to minimize the following performance index (mean square error) [71]:

$$F(x_k) = \frac{1}{N}\sum_{i=1}^{N} e_i(x_k)^2 = \frac{1}{N}\sum_{i=1}^{N} (y_i - y_{mi})^2 \quad (8)$$

where x_k is a vector of current parameters (weights and biases) of the ANN, $e_i(x_k)$ is the training error, y_i is the observed (measured) data of the downstream end water level, y_{mi} is the ANN output (estimation of the downstream end water level developed by the ANN), and N is the total number of samples used to train the ANN.

For the training of the two designed NARX-ANNs, a modified Levenberg-Marquardt optimization algorithm was used, which is a non-linear optimization algorithm that combines the Gauss-Newton method and the Steepest-Descend Gradient method in a single expression and finds the minimum of the function (either linear or nonlinear) over a space of parameters and optimizes the solution [74,75]. This algorithm is characterized by its high convergence speed and reliability, although the complexity of its calculations is greater than in other algorithms [43]. Therefore, the synaptic weights and biases of the NARX-ANNs were adjusted by the algorithm [76]:

$$x_{k+1} = x_k - \left[J^T(x_k)J(x_k) + \lambda_k I\right]^{-1} J^T(x_k)e(x_k) \quad (9)$$

where, J is the Jacobian matrix, $J^T J$ is the Hessian matrix, e is the error vector of ANN between the real output and the estimated ANN output, I represents the identity matrix of appropriate size, k is the number of iterations, and λ_k is the learning coefficient that is automatically updated based on the error at each iteration to secure the convergence.

The FIT crossover performance index was used to quantify the degree of precision of the obtained models. It is defined by the expression [69]:

$$FIT = \left\{1 - \frac{\|y_i - \hat{y}_i\|^2}{\|y_i - mean\ (y_i)\|}\right\} \times 100\% \quad (10)$$

where y_i are the observed (measured) data of the downstream end water level, $\hat{y}_i$ are the estimates of the downstream end water level developed by the model, and $mean(y_i)$ is the means of the downstream end water level.

The simplest procedure to evaluate the quality of the derived ANN model is to train it with the training dataset and validate its performance with different test datasets, e.g., with the validation and testing data sets [77]. The validation data set is used to improve the model performance by fine-tuning the model, while the testing data set is used to provide an unbiased assessment of the final model fit [41].

Validation results of these models with the validation and training data sets are given in Figure 6, which show that the two models adequately describe the dynamic behavior of the water distribution in the canal pool under study with FIT performance indices of 91.0% for the NARX-ANN- based model and 94.0% for the TD-NARX-ANN-based model.

The third model is linear. It was obtained by applying the step response method [69] and is represented by the following transfer function with time delay:

$$G_p(s) = \frac{K_p}{\tau_p s + 1} e^{-Ls} = \frac{0.835}{250s + 1} e^{-300s} \quad (11)$$

where K_p is the static gain, τ_p is the time constant, and L is the time delay.

For linear system identification the Matlab system identification toolbox was used.

Figure 7 illustrates the validation results of the linear model (11) with the estimated nominal parameters with a FIT performance index of 87.0%, which indicates that (11) is a suitable model since models with FIT greater than 80% are considered adequate for process control [69].

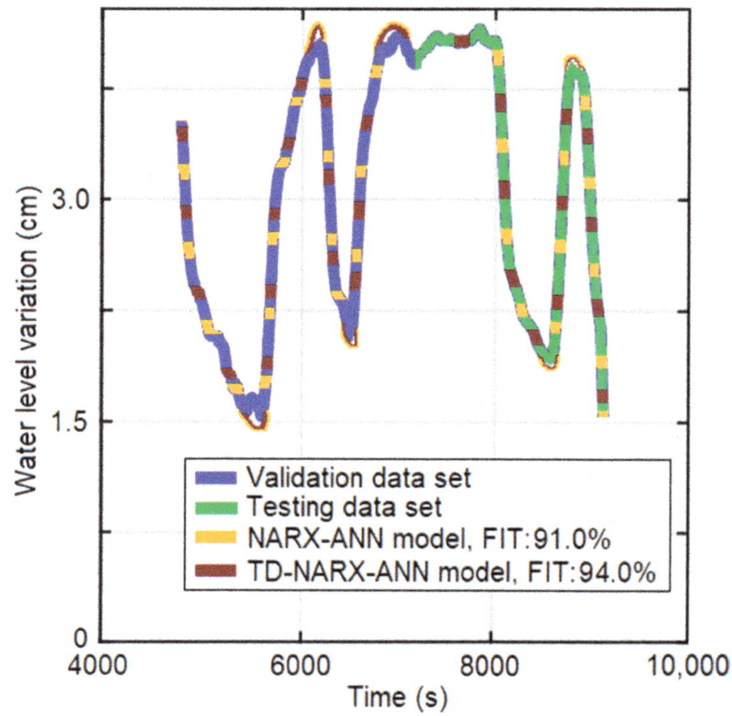

Figure 6. Validation results of the NARX-ANN and TD-NARX-ANN-based models with the validation and testing data sets.

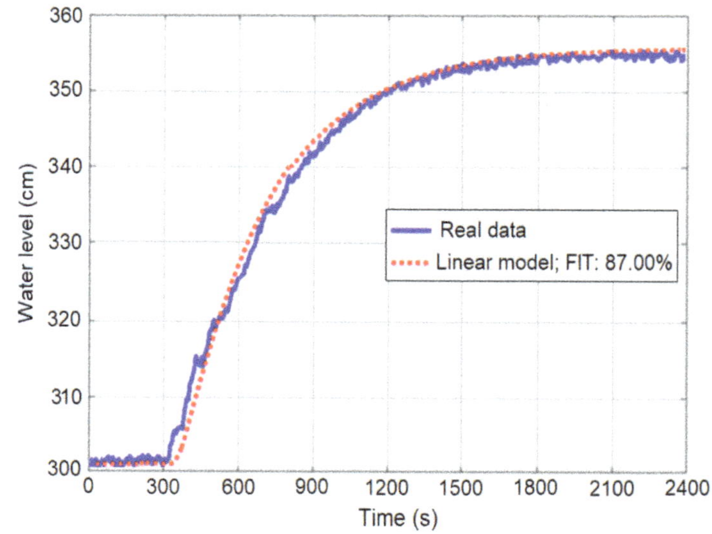

Figure 7. Validation results of the obtained linear mathematical model.

The FIT indexes obtained with the three models demonstrate the better performance of NARX-ANN-based models due to their ability to model the non-linear dynamic behaviors.

3. Design of the NARX-ANN-Based SP Controller

Model (11) shows that the dynamics of the water distribution in our canal pool is characterized by a dominant time delay [28].

Since the time delay reduces the gain crossover frequency, the phase margin, and the speed of response of the control systems, and even can cause the loss of the stability of the closed-loop system [78], the design of a modified SP controller founded on the NARX-ANN-based models previously obtained is proposed for the effective control of water distribution in our canal pool.

The main advantage of the SP is that it removes the real process from the control loop, and it allows then to design the controller from a mathematical model that does not include the process time delay [29].

Therefore, if the mathematical model accurately describes the dynamics of the process to be controlled, the SP eliminates the time delay from the characteristic equation of the closed-loop control system [28].

One of the main problems of the SP controller is the degradation of its performance and robustness when the processes are subjected to the effect of external load disturbances [28,30].

Figure 8 illustrates a diagram of the conventional SP structure, made up of the controller $C(s)$, the real process $P(s)$, and the complete linear mathematical model of the process (integrated by the model without time delay $G_p(s)$ and the time delay e^{-Ls}, which is connected in parallel with the real process. Usually, $C(s)$ is a standard PI or PID controller [21].

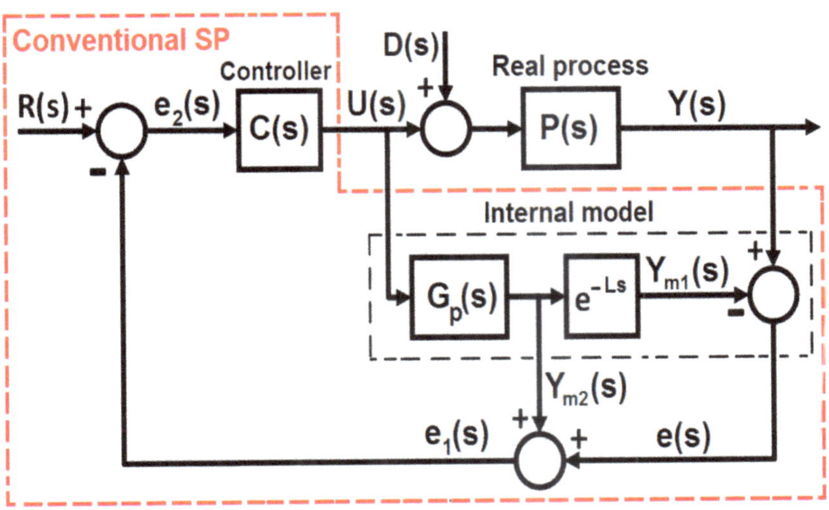

Figure 8. Conventional structure of the SP.

If the mathematical model accurately describes the dynamic behavior of the real process, that is $P(s) = G_p(s)e^{-Ls}$, the output signal $Y_{m2}(s)$ has no time delay and constitutes a prediction of the output signal $Y(s)$ of the real process without time delay [31].

Therefore, based on the mathematical model, it is possible to obtain a prediction of the real process output signal ahead of the time delay. The real process output signal $Y(s)$ is fed back to the controller to remove the negative effect of the load disturbance $D(s)$.

The transfer function of the load disturbance $D(s)$ is the following:

$$G_D(s) = \frac{K_D}{s+1} \tag{12}$$

where K_D is a time varying parameter in the following range: $K_{D\min} < K_D(t) < K_{D\max}$.

If the complete mathematical model $G_p(s)e^{-Ls}$ accurately describes the dynamic behavior of the real process $P(s)$ and there is no load disturbance, the error signal $e(s)$ is zero and the transfer function of the closed-loop control system is represented as:

$$G_{LC}(s) = \frac{C(s)P(s)}{1 + C(s)G_p(s)} = \frac{C(s)G_p(s)e^{-Ls}}{1 + C(s)G_p(s)} \tag{13}$$

If $C(s)$ is a PI controller, the control algorithm is defined by the expression [79]:

$$C(s) = \frac{u(s)}{e_2(s)} = K_C(1 + \frac{1}{T_I s}) \tag{14}$$

where K_C is the proportional gain, T_I is the integral time constant, $e_2(s)$ is the error signal and $u(s)$ is the control signal. Consequently, the controller $C(s)$ is designed for the real process without time delay and the response of the closed-loop control system is time delayed.

The right performance of the SP depends on the exact knowledge of the process model $G_p(s)$ and the time delay e^{-Ls} [28]. Any model uncertainty and/or load disturbance can lead to imprecise time responses of the control system, which can even cause its instability [32].

Our contribution consists of: (1) replacing the linear mathematical model of the conventional SP structure by the non-linear NARX-ANN (3) and the TD-NARX-ANN (4) based models obtained in the previous Section; and (2) adding to the controller output signal $u(t)$ a signal equivalent to the estimate $\hat{D}(t)$ of the load disturbance $D(t)$ (for example, an increase in the lateral discharge $q_1(t)$) in order to reject its negative effect on the control system and, thus, increase the controller performance.

Figure 9 illustrates the diagram of our proposal for the modified SP structure with the non-linear NARX-ANN-based models for the water distribution control in the studied canal pool, where $K(t)$ is a gain tuning block to achieve an adequate estimate $\hat{D}(t)$ of the load disturbance $D(t)$. If there is no load disturbance, $D(t) = 0$, and the error signal is $e(t) \cong 0$. However, when the load disturbance $D(t)$ appears, we have that $e(t) \neq 0$. If the error signal returns to $e(t) \cong 0$ by tuning the $\hat{K}_D(t)$ of the gain tuning block, then the output signal $\hat{D}(t)$ of the gain tuning block is a suitable estimation of the load disturbance $D(t)$.

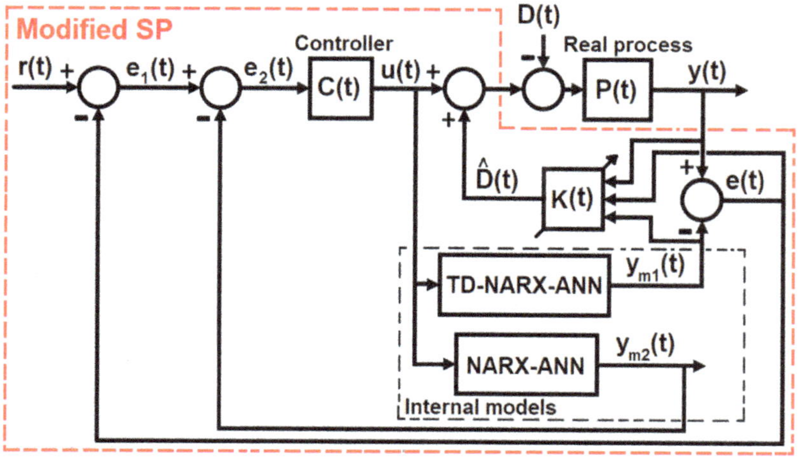

Figure 9. Proposed modified structure of the SP.

The autotuning algorithm of the gain block $K(t)$ is the following:

$$\hat{K}_D(t) = \hat{K}_D(t-1)\frac{y(t-1)}{y_{m1}(t-1)} \tag{15}$$

If $y(t-1) = y_{m1}(t-1)$ the error signal $e(t-1) = 0$ and consequently $\hat{K}_D(t) = \hat{K}_D(t-1)$. But if $y(t-1) \neq y_{m1}(t-1)$ the error signal $e(t-1) \neq 0$ and therefore, the new $\hat{K}_D(t)$ is obtained through multiplying $\hat{K}_D(t-1)$ by a tuning gain obtained as $y(t-1)/y_{m1}(t-1)$.

The linear PI controller in the discrete time domain is represented as [80]:

$$u(t) = K_C e_2(t) + K_I T_s \sum_{j=0}^{t} e_2(j) \tag{16}$$

where $K_I = K_C/T_I$, T_s is the sampling period.

In the proposed modified structure of the SP, the design of the controller $C(t)$ was carried out based on the following non-linear PI control law [46]:

$$u(t) = f\left[e_2(t), T_s \sum_{i=0}^{t} e_2(i)\right] \tag{17}$$

where $f(\cdot)$ is a nonlinear function represented by an ANN, $e_2(t)$ is the control system error, $e_2(t) = r(t) - y(t) + y_{m1}(t) - y_{m2}(t)$, and $r(t)$ is the setpoint.

A multilayer feedforward with tree layers ANN was used to design the nonlinear ANN PI controller. The input layer has two nodes (the system error $e_2(t)$ and the sum of the system error $e_2(i)$), the hidden layer has 5 nodes, and the output layer 1 node. The neuron activation function of the input layer is the linear function $f_i(x) = x$, of the hidden layer is the sigmoid function $f_h(x) = 1/1 + e^{-x}$, and of the output layer is $f_0(x) = x$. Figure 10 illustrates the diagram with the architecture of the designed nonlinear ANN PI controller.

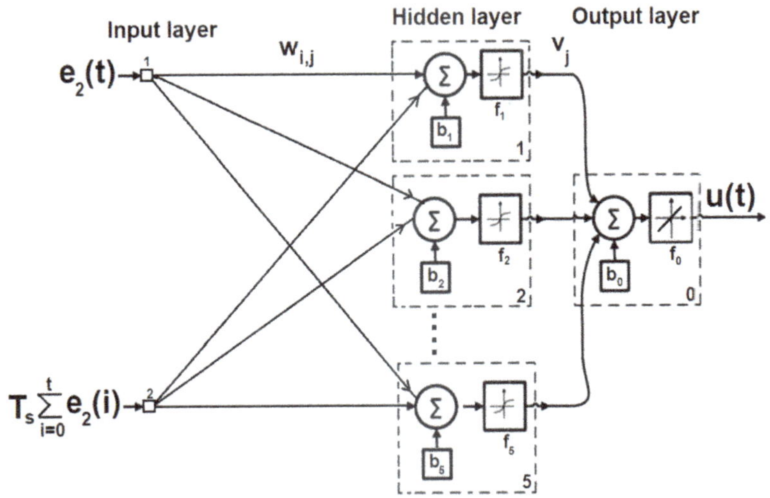

Figure 10. Architecture of the designed nonlinear ANN PI controller.

The data for training this controller was obtained from a discrete-time linear PI controller installed in the canal pool under study.

In addition, using the frequency domain method, a conventional PI controller embedded in a SP structure was designed with specifications: gain crossover frequency $\omega_c = 10^{-3}$ rad/s to achieve a settling time around $T_s = L + \pi/\omega_c \approx 3400$ s and phase margin $\phi_m = 80°$ to obtain a nearly zero overshoot. The controller tuned with these specifications was:

$$C(s) = 0.087(1 + \frac{1}{70.56s}) \tag{18}$$

4. Results and Discussions

For the evaluation of the performance of the modified SP control system with the NARX-ANN-based models, a comparative experimental analysis was carried out with another SP with a conventional structure, that included an internal linear model (13) and did not include the gain tuning block $K(t)$. Both control systems were designed to fulfill the same specifications, considering the following real operation scenarios of our canal pool.

Scenario (1): Experimental performance evaluation of the water distribution control system with both designed controllers when a positive setpoint change $\Delta r(t)$ is applied from 3.10 to 3.55 m (setpoint tracking). Table 1 and Figure 11 illustrate the comparative experimental time responses to that setpoint change yielded by the water distribution control system with the two controllers, where MP is the overshoot, E_{ee} is the steady-state error, T_r is the rise time and T_s is the settling time. Figure 12 shows the experimental comparative output signals of the controllers designed.

Table 1. Experimental comparative results to setpoint change yielded by the water distribution control system with both designed controllers.

Controller	MP (%)	E_{ee}	T_r (s)	T_s (s)
NARX-ANN-based SP	1.0	0	1150	1700
Conventional SP	0	0	2000	3500

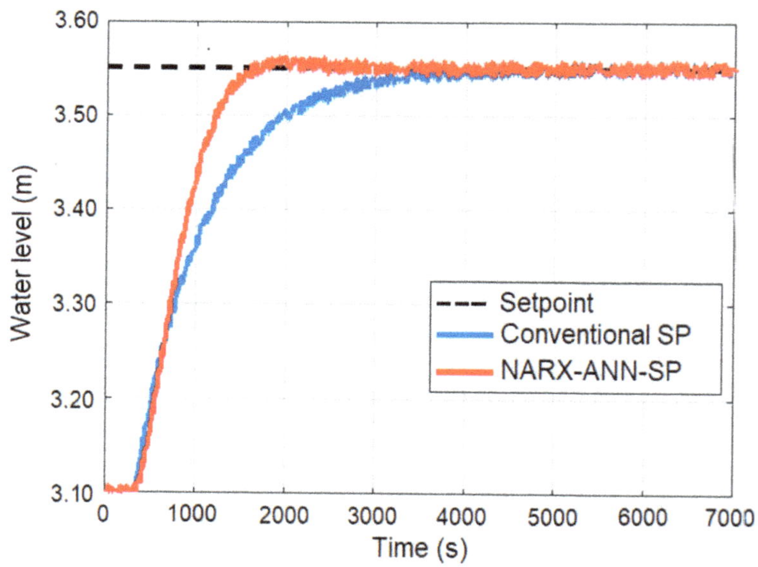

Figure 11. Experimental comparative time responses to a setpoint change of the water distribution control system with both designed controllers.

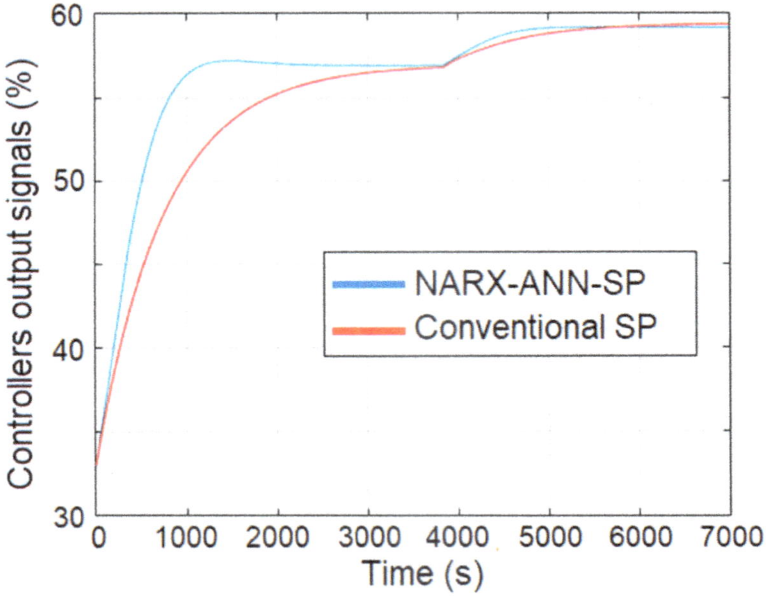

Figure 12. Experimental comparative output signals of the controllers designed.

From Figure 11 and Table 1 it is observed that the control system with the NARX-ANN-based SP controller designed reaches the new reference water level practically without overshoot and with zero steady state error—like the conventional SP—but the rise time and the settling time are approximately the 50% of the times yielded by the conventional SP. After time 3500 s, the responses provided by both control systems have reached the new setpoint with zero steady state error and, therefore, their steady time responses are similar. We highlight that the same responses attained after 3500 s is consequence of the integral terms included in the structures of controller (17) and the PI controller (18), which is a common term in both control systems. However, the main difference is produced in the transient, in which our proposal achieves the final value much faster than the conventional SP controller.

Specifications used to design the conventional PI controller embedded in the SP structure are quite conservative. In theory, much faster responses could be achieved using this control system by tuning controller (18) with higher values of the gain crossover frequency ω_c. However, the value of ω_c used here—though it is not too fast—yields a robust controller (18) that maintains a stable closed-loop response in all the range of operation of the canal, and not only in the nominal operation point at which the controller has been designed. In fact, the highly nonlinear dynamics of the canal makes aggressive linear controllers (controllers that yield very fast damped responses) designed based on a linearized model around an operation point become unstable when the canal operates in a point far away from the linearized one. However, our nonlinear control system based on our TD-NARX-ANN nonlinear model allows a much faster response in the nominal operation point without neither unstabilizing or excessively degrading the closed-loop response of the canal when operating at points far away from nominal. This is caused by the fact that our TD-NARX-ANN nonlinear model captures better than linear model (11) the nonlinear behaviour of our canal pool allowing us to design a controller that takes into account at some extent such nonlinearity.

These results confirm that the NARX-ANN-based SP controller designed improves the management of the water flow through the canal pool and satisfies the users water demand in a shorter time.

Scenario (2): Experimental performance evaluation of the water distribution control system with both designed controllers against the rejection of an external load disturbance $D(t)$. In this case, an increment of 10% in the lateral discharge $q_1(t)$ of the canal pool to satisfy the users water demand was considered as an external load disturbance $D(t)$. Figure 13 illustrates the experimental comparative time responses obtained by the water distribution control system with both designed controllers against the effect of an increment in the lateral discharge ($q_1(t)$) originated in t =3800 s, causing a decrement in the water level of approximately 5.0 cm. The value obtained in the gain tuning block of the designed NARX-ANN-based SP controller was $K(t)$ =0.1.

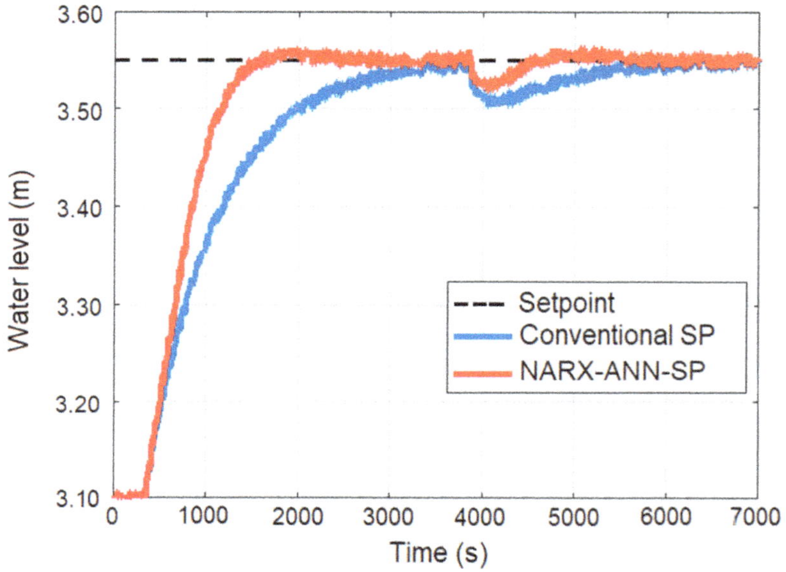

Figure 13. Experimental comparative time responses to the effect of external load disturbance of the water distribution control system with the controllers designed.

These results support the assumption that the best performance of the control system is obtained with the designed NARX-ANN-based SP controller, which makes it possible to fully reject the negative effect of the load disturbance and restore the setpoint water level of 3.55 m in the canal pool in an approximate time of 700 s, while with the conventional SP this load disturbance is fully rejected in an approximate time of 2200 s, which represents an approximate decrement in the rejection time of 70%.

Though conventional SP structures allow very fast responses to setpoint changes of linear systems, their disturbance rejection performance is not good because, though the steady state error caused by a constant disturbance can be driven to zero, the settling time required to achieve this is large. In fact, the basic structure of the SP has been modified several times in the last decades to improve this issue. However, the compromise between the disturbance rejection speed and the closed-loop robustness limits the improvements that can be achieved. An alternative is using the Active Disturbance Rejection Control (ADRC) combined with a SP structure. However, this technique is not well suited to deal with systems with time delay, and the versions that try to copy with this are relatively complicate. We recently developed a modification of the SP structure that included a new feedback term of the error between the real and the predicted outputs [81]. This scheme showed some advantages in rejecting disturbances as well as a better stability robustness than other SP structures. Then, this scheme is a suitable candidate to be used in our canal pool because it yields a faster disturbance rejection, and its improved robustness allows the control system to deal with the nonlinearities of our canal better than other SP schemes.

Moreover, simulations proved that this scheme (in which the linear blocks were substituted by the NARX-ANN blocks in order to better account for the canal nonlinearities) could be further improved by making the gain of that feedback term vary in function of the quotient between the real and the predicted outputs. Then the control scheme implemented in our real canal is [81] but including the feedback term with variable gain that changes according to law (15). Experiments on our canal pool have shown its effectiveness.

These experimental results obtained in the considered real operation scenarios of the studied canal pool show that the designed NARX-ANN-based SP controller outperforms the conventional SP in tracking setpoint changes and rejecting disturbances. Moreover, they show that the proposed NARX-ANN and TD-NARX-ANN models accurately reproduce the canal pool nonlinearities and can be implemented in real time, being more accurate than linearized models while being less computationally involved than more exact models based on the numerical solution of the Saint-Venant equations that cannot be implemented in real time applications.

The NARX-ANN-Based SP controller designed was implemented in the industrial PC installed in the canal pool under study through a program developed in Matlab. This controller communicates via Matlab/OPC (OLE for Process Control) with the SIMATIC S7-300 PLC for real-time data exchange. The OPC technology is based on a Server/Client architecture and provides standards-based communication between PC and PLC.

5. Conclusions

The contribution of this work is a modification of the well-known Smith predictor controller in which the internal linear model has been substituted by the combination of a NARX-ANN-based model and a TD-NARX-ANN-based model, in order to take into account the dynamic nonlinearities in the effective control of water distribution in an irrigation main canal pool.

By application of system identification procedures, NARX-ANN and a TD-NARX-ANN with recurrent architectures have been obtained which describe with high accuracy the non-linear dynamic behavior of the water distribution in the studied canal pool. The NARX-ANN structure with an input layer with 13 memory blocks for the input and output signals, three neurons in the hidden layer and one neuron in the output layer provided the best models performance. A third model which is linear and is represented by a time-delay first order transfer function was obtained using an identification procedure. The validation results of the three models illustrate that the FIT performance indexes of the NARX-ANN-based models are higher than that of the linear model.

In the context of hydraulic engineering, one of the contributions of this paper consists therefore of the fact that accurate nonlinear models based on the NARX-ANN could be obtained for an irrigation main canal pool under different real operation conditions using real-time data and system identification procedures. The results attained are very promising and show that there is great potential for the use of NARX-ANN-based models in depicting the dynamic behavior of irrigation main canal pools. We remark that improving the accuracy of these nonlinear dynamic models has a significant impact on improving the performance of water distribution control systems.

Besides modifying the SP conventional structure by substituting the linear mathematical models by the non-linear NARX-ANN-based models, a gain tuning block of gain $K(t)$ has been added that estimates the load disturbance $\hat{D}(t)$, which is summed to the controller output signal $u(t)$ in order to reject the negative effect of the load disturbance $D(t)$ on the control system and, thus, increase the controller performance.

The experimental results obtained on the real operation scenarios of the studied canal pool illustrate that our modified NARX-ANN-based SP controller improves the control system performance and therefore satisfies the users water demands in time.

The benefits that are obtained with the increment of water distribution control performance in the main irrigation canal pools are reverted to a better management and rational use of the available hydraulic resources, as well as to a greater environment protection due to the reduction of the current operation water losses.

The proposed NARX-ANN-based SP controller constitutes a simple and attractive option for the accurate water distribution control in the irrigation main canal pools and, therefore, it reveals that there is a great potential in the application of the artificial intelligence to design non-linear control systems for this class of hydraulic processes, as well as to design prediction and decision support systems to optimize the operation of irrigation systems.

This research will be expanded in the future to obtain non-linear NARX-ANN-based models of the remaining canal pools of the ZCAMC, and to use these models in the design and implementation of an integral intelligent water distribution control system for the complete irrigation main canal.

Author Contributions: Conceptualization, R.R.-P. and V.F.-B.; methodology, V.F.-B. and R.R.-P.; software, Y.H.-L.; validation, Y.H.-L.; formal analysis, R.R.-P. and V.F.-B.; writing, review and editing, Y.H.-L., V.F.-B. and R.R.-P. All authors have read and agreed to the published version of the manuscript.

Funding: This research received no external funding.

Institutional Review Board Statement: Not applicable.

Informed Consent Statement: Not applicable.

Data Availability Statement: Not applicable.

Acknowledgments: The authors would like to acknowledge the support provided by the Consejería de Educación, Cultura y Deportes de la Junta de Comunidades de Castilla-La Mancha (Spain) with the Project SBPLY/21/180501/000052.

Conflicts of Interest: The authors declare no conflict of interest. The funders had no role in the design of the study; in the collection, analyses, or interpretation of data; in the writing of the manuscript, or in the decision to publish the results.

References

1. Aznar-Sánchez, J.A.; Velasco-Muñoz, J.F.; Belmonte-Ureña, L.J.; Manzano-Agugliaro, F. The worldwide research trends on water ecosystem services. *Ecol. Indic.* **2019**, *99*, 310–323. [CrossRef]
2. Wang, H.; Asefa, T.; Bracciano, D.; Adams, A.; Wanakule, N. Proactive water shortage mitigation integrating system optimization and input uncertainty. *J. Hydrol.* **2019**, *571*, 711–722. [CrossRef]
3. Rijsberman, F.R. Water scarcity: Fact or fiction? *Agric. Water Manag.* **2006**, *80*, 5–22. [CrossRef]
4. Savenije, H.H.G. Water scarcity indicators; the deception of the numbers. *Phys. Chem. Earth* **2000**, *25*, 199–204. [CrossRef]
5. Delgado Munevar, W.G. Gestión y valor económico del recurso hídrico. *Rev. Finanz. Política Econ.* **2015**, *2*, 279–298. [CrossRef]
6. Hernández-López, Y.; Rivas-Perez, R.; Feliu-Batlle, V. Automatic control of water distribution in irrigation systems: Review and challenges. *Rev. Ing. Electrónica Automática Comun.* **2020**, *41*, 80–97.
7. Ali, M.H. *Fundamentals of Irrigation and On-Farm Water Management*; Springer: Berlin, Germany, 2010.
8. Litrico, X.; Fromion, V. *Modeling and Control of Hydrosystems*; Springer: London, UK, 2009.
9. Rivas-Perez, R. Technological Process Control in Main Canals of Irrigation Systems, with Application to Irrigation Systems of Cuba. Ph.D. Thesis, Scientific Research Institute on Land Reclamation and Hydraulic Engineering of Ukrainian Academy of Agricultural Sciences (UkrNIIGIM), Kiev, Ukraine, 1984.
10. Afzal, M.; Battilani, A.; Solimando, D.; Ragab, R. Improving water resources management using different irrigation strategies and water qualities: Field and modelling study. *Agric. Water Manag.* **2016**, *176*, 40–54. [CrossRef]
11. Wahlin, B.; Zimbelman, D. *Canal Automation for Irrigation Systems*; American Society of Civil Engineers: Reston, WV, USA, 2014.
12. Rivas-Perez, R.; Pichuguin, E.D. Synthesis of adaptive control systems of water distribution in main irrigation canals. *Land Reclam. Water Manag.* **1989**, *70*, 66–70.
13. Clemmens, A.J. Canal automation. *Resour. Mag.* **2006**, *13*, 7–8.
14. Rivas-Perez, R. Synthesis of optimal stationary automatic control systems of water distribution in main irrigation canals by means of time delay compensation. *Land Reclam. Water Manag.* **1991**, *74*, 77–82.
15. Lacasta, A.; Morales-Hernandez, M.; Brufau, P.; Garcia-Navarro, P. Simulation of PID control applied to irrigation channels. *Procedia Eng.* **2014**, *70*, 978–987. [CrossRef]
16. Burt, C.M.; Piao, X. Advances in PLC-based irrigation canal automation. *Irrig. Drain.* **2004**, *53*, 29–37. [CrossRef]
17. Arauz, T.; Maestre, J.M.; Tian, X.; Guan, G. Design of PI controllers for irrigation canals based on linear matrix inequalities. *Water* **2020**, *12*, 855. [CrossRef]

18. Bolea, Y.; Puig, V.; Blesa, J. Gain-scheduled Smith predictor PID-based LPV controller for open-flow canal control. *IEEE Trans. Control Syst. Technol.* **2014**, *22*, 468–477. [CrossRef]
19. Feliu-Batlle, V.; Rivas-Perez, R.; Sanchez-Rodriguez, L.; Castillo-García, F.J.; Linarez Saez, A. Robust fractional order PI controller for a main irrigation canal pool. *IFAC Proc. Vol.* **2008**, *41*, 15535–15540. [CrossRef]
20. De Halleux, J.; Prieur, C.; Coron, J.M.; d'Andra Novel, B.; Bastin, G. Boundary feedback control in networks of open channels. *Automatica* **2003**, *39*, 1365–1376. [CrossRef]
21. Astrom, K.; Hagglund, T. *Advanced PID Control*; Pearson Education, S.A.: Madrid, Spain, 2009.
22. Rodriguez Vasquez, J.R.; Rivas-Perez, R.; Sotomayor-Moriano, J.J.; Peran-González, J.R. Advanced control system of the steam pressure in a fire-tube boiler. *IFAC Proc. Vol.* **2008**, *41*, 11028–11033. [CrossRef]
23. Diamantis, M.; Papageorgiou, M.; Kosmatopoulos, E.; Chamilothoris, G. Identification and adaptive control for open channel water flow systems. *Comput.-Aided Civ. Inf.* **2011**, *26*, 464–480. [CrossRef]
24. Zheng, Z.; Wang, Z.; Zhao, J.; Zheng, H. Constrained model predictive control algorithm for cascaded irrigation canals. *J. Irrig. Drain. Eng.* **2019**, *145*, 04019009. [CrossRef]
25. Fele, F.; Maestre, J.M.; Hashemy, S.M.; de la Peña, D.M.; Camacho, E.F. Coalitional model predictive control of an irrigation canal. *J. Process Control* **2014**, *24*, 314–325. [CrossRef]
26. Kovalienko, P.I.; Matzeliuk, E.M.; Rivas-Perez, R. Adaptive control of water distribution in main irrigation canals with variable time delay. *Sci. Res. Hydrotech. Land Reclam.* **1990**, 32–40.
27. Shahdany, S.H.; Taghvaeian, S.; Maestre, J.; Firoozfar, A. Developing a centralized automatic control system to increase flexibility of water delivery within predictable and unpredictable irrigation water demands. *Comput. Electron. Agric.* **2019**, *163*, 104862. [CrossRef]
28. Normey-Rico, J.E.; Camacho, E.F. *Control of Dead-Time Processes*; Springer: London, UK, 2007.
29. Rodríguez, C.; Normey-Rico, J.E.; Guzmán, J.L.; Berenguel, M. On the filtered Smith predictor with feedforward compensation. *J. Process Control* **2016**, *41*, 35–46. [CrossRef]
30. Benitez, I.O.; Rivas, R.; Feliu, V.; Castillo, F.J. Temperature control based on a modified Smith predictor for injectable drug formulations. *IEEE Lat. Am. Trans.* **2015**, *13*, 1041–1047. [CrossRef]
31. Smith, O.J.M. Closer control of loops with dead time. *Chem. Eng. Prog.* **1957**, *53*, 2017–2019.
32. Kao, T.G.; Nguen, M.; Rivas-Perez, R. Adaptive control of a delay plant by using a nonsearching self adjusting systems with a model. *Avtom. Telemekh.* **1988**, *12*, 106–116.
33. Sanz, R.; García, P.; Albertos, P. A generalized Smith predictor for unstable time-delay SISO systems. *ISA Trans.* **2018**, *72*, 197–204. [CrossRef]
34. Castillo-Garcia, F.; Rivas-Perez, R.; Feliu-Batlle, V. Fractional Iα controller combined with a Smith predictor for effective water distribution in a main irrigation canal pool. *IFAC Proc. Vol.* **2008**, *41*, 14558–14563. [CrossRef]
35. Matausek, M.R. A modified Smith predictor for controlling a process with an integrator and long dead-time. *IEEE Trans. Automat. Contr.* **1996**, *41*, 1199–1203. [CrossRef]
36. Shand, M.J. Automatic downstream control systems for irrigation canals. Ph.D. Thesis, University of California, Berkeley, CA, USA, 1971.
37. Deltour, J.L. Application de L'automatique Numérique à la Régulation des Canaux; Proposition D'une Méthodologie D'etude. Thèse de Doctorat, Institut National Polytechnic de Grenoble, Grenoble, France, 1992.
38. Sanfilippo, F. *Application du Régulateur PIR au cas des Canaux à Forte Pente*; DEA de l'Université Claude Bernard: Lyon, France, 1993.
39. Chaudhry, M.H. *Open-Channel Flow*, 2nd ed.; Springer: New York, NY, USA, 2008.
40. Calderon-Valdez, S.N.; Feliu-Batlle, V.; Rivas-Perez, R. Fractional-order mathematical model of an irrigation main canal pool. *Span. J. Agric. Res.* **2015**, *13*, e0212. [CrossRef]
41. Billings, S.A. *Nonlinear System Identification: NARMAX Methods in the Time, Frequency, and Spatio-Temporal Domains*; John Wiley & Sons: New Jersey, NJ, USA, 2013.
42. Chen, S.; Billings, S.; Grant, P. Non-linear system identification using neural networks. *Int. J. Control* **1990**, *51*, 1191–1214. [CrossRef]
43. Medsker, L.R.; Jain, L.C. *Recurrent Neural Networks. Design and Applications*; CRC Press: Boca Raton, FL, USA, 2001.
44. Nikolaou, M.; Hanagandi, V. Control of nonlinear dynamical systems modeled by recurrent neural networks. *AICHE J.* **1993**, *39*, 1890–1894. [CrossRef]
45. Taherdangkoo, R.; Tatomir, A.; Taherdangkoo, M.; Qiu, P.; Sauter, M. Nonlinear autoregressive neural networks to predict hydraulic fracturing fluid leakage into shallow groundwater. *Water* **2020**, *12*, 841. [CrossRef]
46. Tan, Y.; Van Cauwenberghe, A. Nonlinear neural controller with neural Smith predictor. *Neural Process Lett.* **1994**, *1*, 24–27. [CrossRef]
47. Calderon, E.M.; Rivas, R.; Sotomayor, J. Design of neuro-fuzzy controller for control of water distribution in an irrigation main canal. *IEEE Lat. Am. Trans.* **2016**, *14*, 471–476. [CrossRef]
48. Chen, H.; Zouaoui, Z.; Chen, Z. A modified Smith predictive scheme based back-propagation neural network approach for FOPDT processes control. *J. Process Control* **2013**, *23*, 1261–1269. [CrossRef]
49. Choi, H.J.; Jung, S. Neural network-based Smith predictor design for the time-delay in a teleoperated control system. *Artif. Life Robot* **2009**, *14*, 578–583. [CrossRef]

50. Hernández, Y.; Feliu, V.; Rivas, R. Artificial neural network based system identification of an irrigation main canal pool. *IEEE Lat. Am. Trans.* **2017**, *15*, 1595–1600. [CrossRef]
51. Pourfard, A.; Moetamedzadeh, H.; Madoliat, R.; Khanmirza, E. Design of a neural network based predictive controller for natural gas pipelines in transient state. *J. Nat. Gas Sci. Eng.* **2019**, *62*, 275–293. [CrossRef]
52. Huang, J.; Lewis, F.L. Neural-network predictive control for non-linear dynamic systems with time-delay. *IEEE Trans. Neural Netw.* **2003**, *14*, 377–389. [CrossRef]
53. Norgaard, M.; Ravn, O.; Poulsen, N.K.; Hansen, L.K. *Neural Networks for Modelling and Control of Dynamic Systems*; Springer: London, UK, 2000.
54. Narendra, K.S.; Parthasarathy, K. Identification and control of dynamical systems using neural networks. *IEEE Trans. Neural Netw.* **1990**, *1*, 4–27. [CrossRef] [PubMed]
55. Boussaada, Z.; Curea, O.; Remaci, A.; Camblong, H.; Bellaaj, N.M. A nonlinear autoregressive exogenous (NARX) neural network model for the prediction of the daily direct solar radiation. *Energies* **2018**, *11*, 620. [CrossRef]
56. Alsumaiei, A.A. A nonlinear autoregressive modeling approach for forecasting groundwater level fluctuation in urban aquifers. *Water* **2020**, *12*, 820. [CrossRef]
57. Ezzeldin, R.; Hatata, A. Application of NARX neural network model for discharge prediction through lateral orifices. *Alex. Eng. J.* **2018**, *57*, 2991–2998. [CrossRef]
58. Argyropoulos, D.; Paraforos, D.S.; Alex, R.; Griepentrog, H.W.; Müller, J. NARX neural network modelling of mushroom dynamic vapour sorption kinetics. *IFAC-PapersOnLine* **2016**, *49*, 305–310. [CrossRef]
59. Alimohammadi, H.; Alagoz, B.B.; Tepljakov, A.; Vassiljeva, K.; Petlenkov, E. A NARX model reference adaptive control scheme: Improved disturbance rejection fractional-order PID control of an experimental magnetic levitation system. *Algorithms* **2020**, *13*, 201. [CrossRef]
60. Li, H.; Chen, S. A neural-network-based model predictive control scheme for grain dryers. *Dry. Technol.* **2020**, *38*, 1079–1091. [CrossRef]
61. De Giorgi, M.G.; Strafella, L.; Ficarella, A. Neural nonlinear autoregressive model with exogenous input (NARX) for turboshaft aeroengine fuel control unit model. *Aerospace* **2021**, *8*, 206. [CrossRef]
62. Alkhaldi, T.M.N. A novel control strategy for shunt active power filter using NARX Neural Network. *Int. J. Eng. Res. Appl.* **2018**, *8*, 24–30.
63. Manonmani, A.; Thyagarajan, T.; Elango, M.; Sutha, S. Modelling and control of greenhouse system using neural networks. *Trans. Inst. Meas. Control* **2018**, *40*, 918–929. [CrossRef]
64. Medi, B.; Asadbeigi, A. Application of a GA-Optimized NNARX controller to nonlinear chemical and biochemical processes. *Heliyon* **2021**, *7*, e07846. [CrossRef] [PubMed]
65. Katic, K.; Li, R.; Verhaart, J.; Zeiler, W. Neural network based predictive control of personalized heating systems. *Energy Build.* **2018**, *174*, 199–213. [CrossRef]
66. Tan, Y.; Cauwenberghe, A.V. Neural-network-based d-step-ahead predictors for nonlinear systems with time delay. *Eng. Appl. Artif. Intell.* **1999**, *12*, 21–35. [CrossRef]
67. García Láinez, D.; Jiménez Gavilán, P. Propuesta preliminar de caudales de recuperación y conservación del ecosistema del río Zaza aguas abajo de la presa Zaza (Cuba). In Proceedings of the I International Conference on Water and Sustainability, Barcelona-Terrassa, Spain, 26–27 June 2016; pp. 111–113.
68. Miranda Castañeda, L.F. Diagnóstico de la situación de la intrusión marina al sur de Ciego de Ávila. *Ing. Hidráulica Ambient.* **2012**, *33*, 106–118.
69. Ljung, L. *System Identification. Theory for the User*; Prentice-Hall: Upper Saddle River, NJ, USA, 1999.
70. Rivas-Perez, R.; Feliu-Batlle, V.; Castillo-Garcia, F.; Sanchez-Rodriguez, L.; Linares-Saez, A. Robust fractional order controller implemented in the first pool of the Imperial de Aragon main canal. *Tecnol. Cienc. Del Agua* **2014**, *5*, 23–42.
71. Pham, D.T.; Liu, X. *Neural Networks for Identification, Prediction and Control*; Springer: London, UK, 1995.
72. Lin, T.N.; Lee, G.C.; Horne, B.G.; Kung, S.Y. A delay damage model selection algorithm for NARX neural network. *IEEE Trans. Signal Process* **1997**, *45*, 2719–2730.
73. Yazdizadeh, A.; Khorasani, K. Adaptive time delay neural network structures for nonlinear system identification. *Neurocomputing* **2002**, *47*, 207–240. [CrossRef]
74. Azzaro, J.E.; Veiga, R.A. Sliding mode controller with neural network identification. *IEEE. Lat. Am. Trans.* **2015**, *13*, 3754–3757. [CrossRef]
75. Hagan, M.T.; Menhaj, M.B. Training feedforward networks with the Marquardt algorithm. *IEEE Trans. Neural Netw.* **1994**, *5*, 989–993. [CrossRef]
76. Marquardt, D.W. An algorithm for least-squares estimation of nonlinear parameters. *J. Soc. Ind. Appl. Math.* **1963**, *11*, 431–441. [CrossRef]
77. Nelles, O. *Nonlinear System Identification from Classical Approaches to Neural Networks, Fuzzy Models, and Gaussian Processes*, 2nd ed.; Springer: Berlin, Germany, 2001.
78. Feliu-Batlle, V.; Rivas-Perez, R.; Castillo-García, F. Design of a PIα controller for the robust control of the steam pressure in the steam drum of a bagasse-fired boiler. *IEEE Access* **2021**, *9*, 95123–95134. [CrossRef]
79. Ogata, K. *Modern Control Engineering*, 5th ed.; Prentice Hall: New Jersey, NJ, USA, 2010.

80. Ogata, K. *Discrete Time Control Systems*, 2nd ed.; Prentice Hall: New Jersey, NJ, USA, 1995.
81. Feliu-Batlle, V.; Rivas-Perez, R. Control of the temperature in a petroleum refinery heating furnace based on a robust modified Smith predictor. *ISA Trans.* **2021**, *112*, 251–270. [CrossRef] [PubMed]

MDPI
St. Alban-Anlage 66
4052 Basel
Switzerland
Tel. +41 61 683 77 34
Fax +41 61 302 89 18
www.mdpi.com

Applied Sciences Editorial Office
E-mail: applsci@mdpi.com
www.mdpi.com/journal/applsci

www.ingramcontent.com/pod-product-compliance
Lightning Source LLC
Chambersburg PA
CBHW041953260926
51850CB00019B/2687
9783036568089